- -

NAVIES OF THE WORLD
SPECIAL PUBLICATION - ARCHIVAL EDITION

- -

A VISUAL HISTORY & REFERENCE FOR SUBMARINE INSIGNIA OF THE WORLD

DOCUMENT NO: 979-8-218-86695-2
REFERENCE CODE: HISTORICAL ISSUE
ISSUING BODY: SUBMARINE INSIGNIA RESEARCH DIVISION
AUTHORS:
CHRISTOPHER WALKER
LEE LUKAS

PREPARED UNDER THE AUTHORITY OF:
SUB INSIGNIAPEDIA PUBLISHING
2026

NOTE: THIS DOCUMENT IS PRESENTED FOR HISTORICAL AND EDUCATIONAL PURPOSES. IT IS NOT AN OFFICIAL U.S. NAVY DOCUMENT.

STAMP: [DECLASSIFIED]

- -

FOR HISTORICAL RECORD - ARCHIVAL COPY

- -

COPYRIGHT

Visit the authors' website at www.subinsigniapedia.com

ISBN: 979-8-218-86695-2 (print)

The contents of this book are derived from the most accurate and reliable resources available to the authors through their research and experiences. These contents do not alleviate the reader of the responsibility for conducting their own research. As with any collectible, the responsibility for verifying authenticity lies with the buyer.

Published by Sub Insigniapedia Publishing

1st Edition 2026

TABLE OF CONTENTS

DEDICATION

Chris Walker

To my wife, Alison, thank you for your love, your patience, and your constant support. You've stood by me without ever questioning the time or space this hobby has taken, and you've always seen each badge not merely as a piece of metal on the wall, but as something meaningful simply because it means something to me. Your support has been the foundation of it all.

Lee Lukas

To my wife Stacey, who not only tolerates, but actually graciously supports my obsession with/addiction to collecting submarine insignia. And to my children, Matt, Emily, and Audrey for occasionally taking the time to listen to my ramblings on the subject without rolling their eyes. Your love and support mean more than I can adequately express.

To our fellow collectors and collaborators – as former submariners we both started our collections with the simple premise of compiling submarine badges from as many countries as possible. After meeting and consulting with many of you, our focus was changed to learning as much as possible about the history and facts surrounding these amazing items. Your wisdom, insight and incredible generosity with your most valuable commodity – time – have been the impetus to create this volume. We are forever in your debt for welcoming us with open arms into this fraternity.

ACKNOWLEDGMENTS

Those of us who have served aboard submarines are acutely aware of the "Brotherhood of the Phin"… the fraternity of men and women who, in times of peace and conflict, operated a multi-million/billion dollar machine in a hostile natural environment. There are very few other jobs where personnel live and work in such close proximity while dealing with potentially life-threatening situations. Just as pressure forms diamonds, it also forms bonds that last a lifetime.

A similar fraternity exists for those in the collecting community. It is a community that, during its finest hour, is based on mutual respect, the sharing of knowledge, friendly competition, and the thrill of the hunt.

The creation of this publication would not have been possible without the help, support, and encouragement of our "Sea Dads"… those icons of the submarine insignia collecting community who developed a deep knowledge and understanding of submarine insignia history, much of it pre-internet and pre-email. The authors owe a huge debt to the following groups for igniting our curiosity, and fueling our passion:

First, to the pioneers who made the initial notable efforts to capture the many details associated with the submarine insignia of the world: LCDR (SS) Allen Phillips, USN, Pete Prichard, LCDR Maitland Thornton, RN, Capt. Gustavo Conde, Argentine Navy, David Jones, CDR (SS) Steve Jones, USN, the creator of the first comprehensive submarine insignia website, and William Crosby, who made significant contributions to almost all of the above. Truly a herculean task accomplished by each of them through massive letter-writing and e-mail campaigns, travel, and untold hours on the phone. Their efforts paved the way and set a standard we can only hope to approach.

The following individuals have earned our undying admiration, respect, and gratitude for not only enduring hundreds (maybe thousands!) of questions, but also selflessly sharing their time, information, knowledge and unique insights into submarine insignia collecting:

Mike Brookins
Maurice (Vic) Borg
William Crosby
David Jones
Steve Jones
Sergei Kazeka
Mike (Lee) Lacey
Hans Lauritzen
Torkil Nissen
David John Payne
Owen Pate
Peter Van de Put

Additionally, there are people who provided invaluable assistance for individual or a few countries, reviewing the chapters for accuracy, and in many cases graciously sharing their research and contacts with us to contribute to a more accurate end product:

Ivan Abraham
Massimo Agati
T.J. Carter
Alf Chern
Eric Fraser
Eric Gragnac
Piero Fagandini
Pär Flyghed
Leonid Kanterman
Rick Kennedy
Akira Komiya
Nick Komiya
Walter Kryskiw
Adam Tomasz Krzykowski
John Lamont
Jovan Mara
Alberto Menichetti
Bill Miller
George Quigley
Zino Stef
Brooke Walker
Junhao Gregory Wong

No doubt there are people along the way who have aided us in our quest that are not named here. These are omissions of memory and not of intent. There are also a few sources who provided valuable aid but wish to remain anonymous. This was truly a community effort, and we sincerely appreciate the contributions of everyone to this project.

With so many helpful contributions, it was inevitable that we would find some conflicting information along the way. When that occurred, we presented the information that made the most sense to us based on the best available information at the time. Any errors are ours, and ours alone, and we eagerly anticipate feedback from the collecting and historical communities to allow us to make necessary revisions in a future updated version of this guide.

PREFACE

Treasure hunting. From the quest for the Holy Grail, to the search for El Dorado, seeking some form of treasure has captured the human imagination for thousands of years. For the past 100+ years, submarine insignia collecting has done the same, albeit on a MUCH smaller scale.

What started off as a show of camaraderie where submariners of different countries would exchange badges as tokens of good will when meeting in foreign ports, has turned into an international bazaar with collectors in all corners of the globe vying for bragging rights by obtaining badges to fill holes in their collections. The badges once only seen on a submariner's chest are now found in a variety of locations, with many found in online auctions.

How did we get here? The first official submarine insignia were two patches issued by the French Navy in 1906, followed a few years later by the Imperial Russian Navy issuing the first metal insignia in early 1909, Japan later that same year, and then Austria-Hungary in 1910. These badges, and many of those that followed in the next 20 years were true works of art, and the pride of those that earned the right to wear them. Because of the extremely limited population of sailors participating in the infancy of submarines (few countries with submarines, limited crew sizes, and a limited inventory of submarines), the numbers of pre-1930s submarine insignia issued were relatively small.

As more and more countries developed or purchased submarine technology, the number of countries with qualification/designation badges grew. From the late 1800s to today, at least 56 different countries have had official naval submarines, the majority of them with submarine insignia of some form.

Many of these countries have had multiple variations of their insignia. From different designs, manufacturers, and levels of qualification to those required for different uniform standards, the known variations for a single country can easily run into double digits. While some countries have kept it very simple with only one or two variants, others have taken the opposite approach. The leaders of this group are France, Italy, Russia/USSR, and the United States who all easily have over 100 different known variations of submarine-related insignia. Please note the use of the phrase, "known variations"… every year new, previously unknown variations of insignia come to light in the collecting community.

So, with that information in mind, one can easily see that applying the Pokémon philosophy of "Collect 'em All" to submarine insignia can be a rather daunting task. Our goal in creating this book is to provide the collecting community with a detailed and an as historically accurate as possible guideline that builds upon existing knowledge, corrects some previously existing misconceptions, and expands the scope to also cover many of the known variations in design and manufacturer. While we believe that has been accomplished in these 234 pages, we are aware that we are still scratching the surface in some regards.

From the inception of this project, we have maintained a focus on providing information that is as fact-based as possible. However, with huge gaps in the available records from the different Navies and badge manufacturers around the world, it can be difficult to ascertain the specific timeframe certain designs were in production, or the quantities made. Similarly, when researching uniform regulations or other standards guiding the issuance or wear of submarine insignia, determining the duration certain insignia were in use or required to be worn can be exasperating at times.

So, while this goal has been challenged by the passage of time, we have made every effort to remain true to the values of "Just the facts ma'am". When information is presented as fact, it is because it is backed up by documentation that has been verified by official and/or trusted sources to the best of our abilities. Any other information will have familiar literary caveats such as, "Based on the best available information", "according to sources", and "it is generally accepted." There are some very interesting theories and legends out there that will not be presented here as facts, as we have tried to present only histories that have at least two corroborating sources. As those who study the past understand, history is written by the winners. We have done our best to cut through the fog of the past to uncover the rest of the story – ultimately hoping for the truth.

In compiling such a large trove of information, we quickly discovered that it is unlikely we will succeed in presenting this information in a manner that will please everyone. We choose to be very pleased that not everyone will agree with all of the information presented in this book. Differences in opinion spark conversation and debate… the building blocks of progress and understanding. Our deepest hope is that this book will generate intense discussions that will result in bringing more verifiable historical documents and facts to the surface, filling holes in the historical record that we were unable to. To that end, the authors have established a clearing house for submitting theories, questions, stories, legends, and cold hard facts. Information can be submitted to subinsigniapedia@gmail.com, and authors of leads resulting in modifications will be acknowledged in the next edition of this guide, and receive the satisfaction and bragging rights that come with being able to tell the submarine insignia collecting community, "I toldja so!"

CAVEATS, CONVENTIONS & CLARIFICATIONS

In no particular order:

For identifying badges, we have chosen to build upon the naming convention established by Pete Prichard, as this has become the standard throughout the collecting community. Where it was necessary, we have made alterations/modifications to support verification of timelines or country provenance, and/or inclusion of new information. For example, Prichard's ITL-3 is now ITL-1 based on it being verified as the first Italian submarine distinctive Trades badge from 1918, pre-dating his ITL-1 & ITL-2 (now ITL-2 & ITL-3) issued in 1924.

The terms used to describe the construction and design of badges will be those of a layperson. We are not heraldic experts, badge manufacturers, nor designers. If we get a description incorrect, please let us know and we will correct it in the next edition.

In general, we will not be showing badges that are widely known to be fakes, such as the U.S. badge with the skull and crossbones. However, there are some badges that have become so accepted by the collecting community as genuine submarine qualification insignia that any comments to the contrary are met with scorn and disbelief. Some of these badges will be shown just for the purpose of debunking their validity. We do not take this debunking lightly, and only do so after corroboration of the facts by Naval authorities of the country in question.

We will also not be discussing the various "small unit" badges discussed in other books. These units were typically made up of personnel in dive gear operating "wet" submersibles. There are a few exceptions to this such as the German WWII *Seehund*-class mini subs, but we will stand by their absence until convinced to include them in a future edition of this guide.

The U.S. Combat Patrol badge and the U.S. and United Kingdom Fleet Ballistic Missile Submarine badges are considered to be qualification insignia for the purposes of this book and will be shown as such. Similarly, "patrol" examples of badges from Germany, Italy, and the Soviet Union will be shown.

Boat badges: There are some countries that have beautiful badges/pins that were created for specific submarines. We like to call these "boat badges". Countries like Italy, France, and Russia have literally dozens of examples of these. For these three countries in particular, there already exist wonderful books/references that detail the history and beauty of these badges/pins... but they are NOT qualification nor designation insignia, so we will not be showing many of them here. However, there are several countries that have a small number of boat badges that have never made it into a book. So, at the risk of being ever so slightly hypocritical, because some of these are nicely made and, in some cases, incredibly rare, we have made a few exceptions to the "No boat badges" rule, and those will be appropriately highlighted.

As previously mentioned, there are multiple variations of many badges in this book. While we have done our level best to include as many as reasonably possible, it is inevitable that some will be left out… many on purpose (due to lack of available space) and some due to our lack of knowledge of their existence. We encourage collectors who notice missing variants to send us photos for possible inclusion in the next edition.

While "officially sanctioned" badges are the bulk of those shown here, there are many badges that become accepted as "official" without ever being sanctioned by a Naval authority. We have attempted to notate most situations where this has occurred.

The majority of the photos in the book are from the collections of the authors. For those badges we do not possess, we relied upon the kindness of the previously mentioned team of collectors to supply us with photos of badges in their collections. In rare cases, though evidence of some badges exists, the authors were unable to find physical examples within our sphere of known collectors and experts. So, for these, we have either inserted a sketch, or given them a placeholder identifying number, but followed by an asterisk (*) signifying "Not Pictured".

COUNTRIES WITHOUT KNOWN SUBMARINE INSIGNIA

Albania

Albania established a submarine unit in 1958 with two *Whiskey*-class submarines from the Soviet Union. Several years later, Albania pulled out of the Warsaw Pact and ended up with four of the submarines, whose ownership continued to be disputed with the Soviet Union for years. China stepped in to help the Albanians operate the submarines, but by the 1980s Chinese assistance had evaporated and the submarines fell into disrepair, becoming basically inoperable shells by the end of the decade.

There is no verifiable evidence of any Albanian submarine distinctives/qualification badges that existed during any of the phases of the Albanian submarine history. There is a patch with a silver starboard-facing submarine profile on a black background (ALB-UA) that, according to sources, was supplied to Albanian submariners by a collector. The authors have been unable to find any official records regarding any Albanian submarine insignia, so, if used at all, the only likely scenario is that the patch was used in an unauthorized capacity. Known copies of this badge exist.

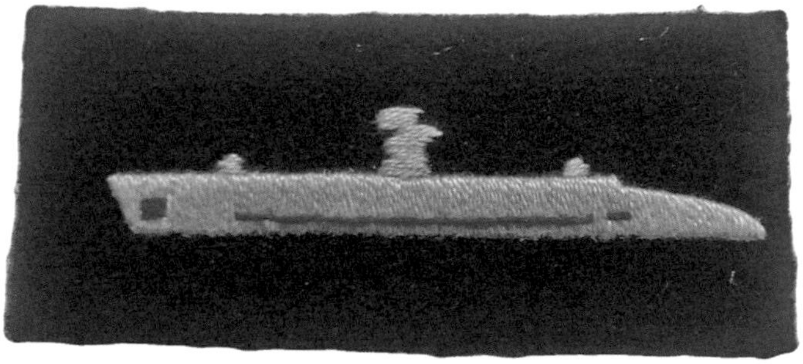

ALB-UA

Azerbaijan

After decades of Soviet rule, the independent Azerbaijani navy was established in 1992, a year after the fall of the Soviet Union. Though independent, the Azerbaijani navy still heavily leans on Russia for support. Azerbaijan owns and operates four Triton midget submarines acquired from Russia that are piloted and operated by personnel in wet suits. There is no knowledge of submarine insignia associated with these submariners.

China

Mao Zedong attached such importance to a Chinese submarine force that the Chinese Communist Party authorized the formation of a submarine force in June 1949, four months prior to the official founding of the People's Republic of China (PRC). Several hundred officers were selected as the core for a new submarine fleet of the People's Liberation Army Navy (PLAN) in 1951 and sent to study and train aboard Soviet submarines, but it was not until 1953 that China received its first submarines from the USSR.

During the very early years of the PLAN's submarine force, a close bond was maintained with the Soviet navy. While there is no known official PLAN submarine insignia, there is a badge that has been identified as a 1950's period Soviet Submarine Commander badge with the People's Liberation Army's (PLA) symbol in the center, the numbers "8" and "1" in a red star, symbolizing creation of the PLA on 1 August 1927 (CHN-1-UA). The veracity of this badge has not been confirmed, however, we have included it here because the possibility exists for such badges to have been created, and if genuine, it is an incredible piece of history. While Sino-Soviet relations soured beginning in the late 1950s, the PLAN submarine force continued to grow, eventually becoming self-sustaining and not relying on other countries for submarines.

According to multiple sources, there has never been an official PLAN submarine distinctive of any type. In the 1990s a badge appeared purporting to be a PLAN submarine badge. This bronze-colored badge has a bow-on port-leaning submarine profile in the center of a wreath topped by a five-pointed star and flanked by stylized waves, with the Latin letters "HGTZH" underneath, and Chinese characters on the back that translate to "PRC Navy" (CHN-2-UA). This badge is likely fake, and only shown here to emphasize that there is no evidence it was ever authorized for wear by PLAN submariners.

CHN-1-UA

CHN-2-UA

Croatia

The lone submarine operated by the Croatian navy was the *Una*-class midget submarine *Velebit*, captured from Yugoslavia in 1991. It was overhauled and put into operation in 1996 and was in use until being decommissioned in 2006. Though veteran's badges exist, there is no evidence of a Croatian submarine distinctive existing during the submarine's use. Fake badges purporting to be either Croatian or for the post-1992 Federal Republic of Yugoslavia (FRY) exist, similar in all aspects to the Yugoslavian Senior Submariner badges except with the FRY coat of arms at the top instead of the red star.

Estonia

Estonia only had a submarine force for three years – from 1937-1940. According to official sources, during this time, submariners wore the same insignia all members of the navy wore, and no special submarine distinctive was created. There are several examples of commemorative and fund-raising badges that exist, but none of these had any official standing.

Finland

After achieving independence in 1917, Finland quickly identified strategic military needs and saw the value added by establishing a submarine force, beginning construction on submarines in 1925. The Finnish submarine force was officially born with the commissioning of the first of three *Vetehinen*-class submarines in October of 1930. These were followed a few years later by two smaller submarines. These five submarines were all in service during WWII, but following the war Finland was no longer allowed to operate submarines, and all were removed from service.

There is no evidence that an official submarine distinctive was ever used by Finnish submariners. A badge (possibly a crest) with the date of 1929 on it and the words "Suk. Vene Koul.", short for *Sukellus Vene Koulu,* or Diving Boat School, is known, with an example residing in the Military Museum in Helsinki (FIN-SCHL-1). There are so-called "bottom plates" or unfinished versions of this badge that have surfaced as well (FIN-SCHL-1UF). Very few examples of either of these are known to exist.

FIN-SCHL-1

FIN-SCHL-1UF

A badge by the same Finnish maker of the submarine school badge exists that has an older model stylized port-facing submarine profile surmounting a three-bladed propeller, encircled with a ring having a blue enamel groove (FIN-??). Again, there is no evidence this was ever considered to be a submarine distinctive authorized by naval authorities, however it is possible that it was a badge that crew members had fabricated to represent a certain submarine. It is a very high-quality badge so it is also possible it could have had a more official status but its status, official or otherwise, is unknown at this time.

FIN-??

There was a submarine veteran's stick pin used for many years by the WWII veterans that is a starboard-facing submarine bow-on into the waves, enclosed in a ring with "SuvLv" on the bottom (FIN-VET). According to sources, this pin was initially distributed in 1960 and was silver plated. Years later when the supply of pins ran out, 100 more were ordered, but when they arrived the new batch was gold. This is not an indication of status, just a manufacturing mistake.

FIN-VET

Finland has expressed interest in obtaining new submarines at various points over the past 20 years, and that effort appears to be recently reinvigorated, so it is possible that at some point we may see official Finnish submarine badges.

Latvia

Latvia purchased two French-built submarines and commissioned them in 1927. These were operational for over a decade and were in use when the Baltic States were annexed by the Soviet Union in 1940. However, both were laid up for repairs when they were scuttled in the port in 1941 to avoid capture during the pending German invasion. There is no evidence of any distinctive insignia worn by the Latvian submariners other than cap tallies for the *Spidola* and the *Ronis*, however, like Estonia, there are several commemorative and fund-raising badges that exist.

North Korea

In the realm of secretive militaries, North Korea is likely the winner in the category of keeping public knowledge of their abilities hidden. Virtually nothing is publicly available regarding information associated with things like potential submarine insignia. So, it almost goes without saying that there is no confirmed information regarding an official (or otherwise) North Korean submarine insignia… but we will say it anyway! This includes the patch created by Steve Jones many years ago, complete with fake Korean symbols, that he will be the first to tell you is not genuine.

Syria

Syria supposedly operated three *Whiskey*-class Soviet submarines in the 1960s. They then obtained three *Romeo*-class Soviet submarines in the mid-1980s, and these operated several years before being decommissioned and scrapped before the end of the 1990s. There is no reliable information on any submarine distinctives being used by the Syrian navy during either of these periods.

Ukraine

The Ukrainian submarine force began in 1997 with the acquisition of *Zaporizhzhia*, a *Foxtrot*-class submarine from the Russian navy. She underwent extensive repairs and was in a continual state of overhaul until she finally put to sea in 2012. She was seized by the Russian navy in 2014, and currently remains pier-side at a Russian facility.

According to Ukrainian sailors assigned to *Zaporizhzhia*, at no point in her 17 years with the Ukrainian navy was there ever any submarine distinctive authorized for wear by the crew other than winter and summer unit patches. The captain wore a Soviet "Submarine Commander" badge, earned prior to the fall of the Iron Curtain. There is a drawing purporting to authorize a proposed "Submarine Commander badge", supposedly issued in 2012 (UKR-PROP-1), however no badge was issued prior to the boat being taken over by the Russians in 2014. The badges currently on the market claiming to be Ukrainian submarine distinctives are the work of veterans of *Zaporizhzhia*, but they had no official affiliation with the ship itself.

UKR-PROP-1

PROPOSED BADGES FOR COUNTRIES WITHOUT SUBMARINES

The badges in this section are for countries that do not have submarines, but have had interest in having them, to the point of having a badge created… or perhaps more accurately, created for them by badge manufacturers and designers hoping to be on the leading edge of the country's submarine program development.

Bolivia

Bolivia has flirted with the idea of having submarines since the 1960s, but the plans have never come to fruition. A badge was proposed to the Bolivian Navy by Herbert Booker, a noted designer of badges, whose sketches have been in multiple books. The proposed Bolivian badge is a gilt profile of a starboard-facing submarine profile on a flat gold circle. The actual proposed design was in sketch only, but badges have been manufactured for the collecting community based on the sketch (BOL-PROP-1).

BOL-PROP-1

Nigeria

In the early 2000s, Nigeria was considering purchasing a submarine, so representatives in the Nigerian embassy in the United States asked a U.S. manufacturer to design a proposed submarine badge for them. The manufacturer (KREW) used the U.S. Navy Medical dolphin badge as a template and replaced the center oval and acorn with a "seal" that had a fouled naval anchor with an eagle above it, the words, "Nigerian Navy" at the top of the oval, and "Forward Together" on the bottom. This badge had a slightly convex surface, and approximately 100 of these were made (NIG-PROP-1). Later, reproductions were made of the badge for the collecting community with the exact same design, but the surface of the seal is flat (NIG-PROP-1R). The reproduction is smaller (68mm long) and hollow back, while the original (70mm long) is solid back.

NIG-PROP-1

NIG-PROP-1R

Philippines

The Philippine government has expressed interest in acquiring submarines in one form or another for over a decade, but negotiations began in earnest in 2019. These efforts proved to be serious enough that the company Santi was commissioned to provide a badge design. Badges were manufactured in 2020 that have a profile of a diagonally starboard-facing submarine flanked by two outward-facing sharks riding stylized waves and a wreath with a star in the center arching over the submarine's sail. Examples exist in copper (PHL-PROP-1-C*) and a silver version with the same basic design, but a gold submarine (PHL-PROP-1-S). A newer version seen in 2024 is copper-colored, but without the wreath (PHL-PROP-2-C). Sources report the copper-colored version is considered temporary and the eventual plan is to have these in gold for officers and silver for enlisted personnel. With completion of this acquisition at least six to ten years away, it may be quite some time before the Philippines have an approved qualification insignia.

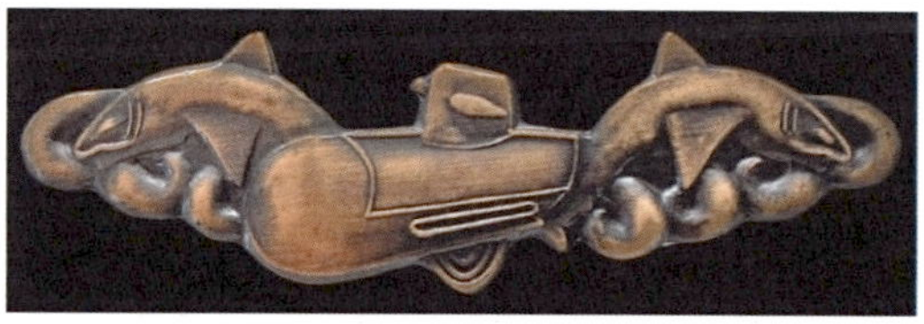

PHI-PROP-2-C

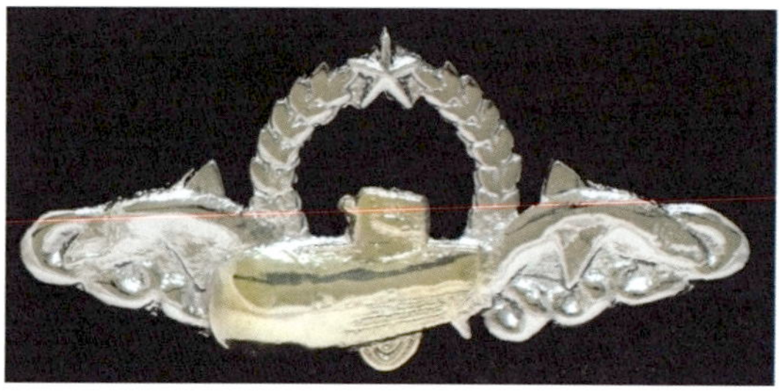

PHI-PROP-1-S

Saudia Arabia

There are few badges that evoke emotion and conflicting opinions like this Saudi Arabian badge. Many experienced collectors are convinced it is a Combat Swimmer/Deep Sea Diver badge, while just as many believe it to be a proposed submarine badge. We include it here because we have no proof one way or the other, but it is conceivable that both possibilities can be true. If the divers operating Saudi midget submarines were the intended recipients, then this badge could fit both bills. Supposedly manufactured in the early 1980s, in true Saudi fashion, it is large and impressive. Gilt in color, two inward-facing stylized dolphins face an unfouled naval anchor, above which are the crossed scimitars and palm tree of the Saudi Arabian national emblem, with the royal crown at the very top and six lightning bolts emanating in an arc from either side of the national emblem (SAU-PROP-1.1). A pierced version is also known (SAU-PROP-1.2). While the authors found no evidence validating this badge, full-size and mini versions exist in varying levels of quality due to likely copies.

SAU-PROP-1.1

SAU-PROP-1.2

ALGERIA (ALG)

While the capabilities of a nation's submarine force are typically cloaked in secrecy, some nations do this better than others. Algeria is one of those countries. The Algerian submarine force was born in 1983 with the purchase of two *Romeo*-class submarines from Russia and expanded upon with the purchase of two *Kilo*-class submarines in 1987. Because the authors could find so little information about the small Algerian submarine force, it is not known if there were any submarine distinctives created and issued during that period.

While the dates and circumstances of initial issuance for the first submarine insignia are not currently fully understood, it was not until around 2010-2012 that photographs emerged of potential Algerian badges that eventually became verified. The first known badge is two-part construction with a gold port-facing stylized submarine profile with the Algerian flag on a very prominent sail. This submarine is mounted on a silver shield comprised of a wreath that has laurel and oak leaves on the sides, with an interpretation of the Algerian Naval emblem in the center. An unusual feature of this badge is that the details of the shield are continued on the embossments on the front of the submarine. This is a clutch back badge with both pieces having a single individual pin (ALG-1). Source report this badge was worn by the Commanding Officer and Engineering Officer of the submarine on the left breast of their uniform and has a maker's mark on the back. A smaller sized, possibly mess dress, version of this badge has been identified, with all details exactly the same, except the submarine is 52mm long (ALG-1m*) and does not appear to have a maker's mark. The purpose of the smaller badge has not yet been verified by the authors.

ALG-1

A similar badge is likely for the other officers and possibly also senior enlisted personnel. This is also a port-facing stylized submarine that is the exact same dimensions and design as the submarine used for ALG-1, with the only difference being that it is a silver submarine and not gold. This is paired with the same silver shield as the ALG-1 (ALG-2). While currently unverified, it is possible a mess dress version exists for this badge as well (ALG-2m*). Because of the two-part construction of these badges, examples are known to exist of only the submarine because the shield is missing. This is a common issue with many two-part construction badges.

ALG-2

In the early 2020s a third style badge was identified. This one is a gilt metal stylized submarine that is 85mm long with multiple limber holes shown, and a very low-profile sail bearing the Algerian flag. This badge also has a single clutch pin (ALG-3). Neither the purpose nor the reason for which this badge is awarded is currently known.

ALG-3

In 2023 a new style badge surfaced purporting to be the new Algerian submarine qualification badge. This gilt badge features a port-facing submarine profile laying across the flukes of a very wide anchor, with the anchor chain wrapping behind the submarine and over the front right anchor arm (ALG-4?*). This badge has three screw connection points on the back. The authors do not have permission to show the photo of this badge. While it was supposedly presented by a member of the Algerian submarine force, no corroborating evidence of it serving as the new Algerian submarine qualification insignia has yet been identified.

ARGENTINA (ARG)

Though experimenting with submarines as early as 1907, the Argentine Navy did not establish a formal submarine service until it commissioned three Italian-built submarines in 1932. Shortly after the submarines arrived at their home base on 3 September 1933, a qualification badge was implemented. On 16 Dec 1933, the President of Argentina issued Decree No. 32.871, officially sanctioning and providing design guidance for submarine insignias that would be issued to the personnel assigned to their new submarine force. As a source of great national pride, it was unusual, but fitting, that the President himself would issue this decree.

The decree modified Article 46 for submarine personnel to provide for a gilt (gold) metal "senior personnel" badge for officers consisting of a port facing submarine profile featuring a conning tower with two periscopes, and an oval in the center encircled by a rope. This oval contains one of the national symbols of Argentina featured on its flag, an enamel Sun of May, on horizontal stripes in the national colors of light blue and white (ARG-1).

ARG-1

It also provided for a gilt metal "junior personnel" badge (that appears more bronze than gilt) for the enlisted ratings personnel with the same measurements as the officer's badge, the difference being an engraved naval anchor in the middle of the oval at the center of the badge without the enamel (ARG-2). Both badges had horizontal pin attachments and were to be worn "on the left side of the outer garment at the height of the nipple". Two versions of the ARG-2 badge are known, one with a fully horizontal anchor stock (ARG-2.1), and the other with a diagonal anchor stock (ARG-2.2). The significance of the difference is unknown, and it is likely just due to difference in manufacturers. There have been modern restrikes of the ARG-1 and ARG-2 badges made that look exactly like the originals on the front, so provenance is key. Sources indicate it is possible that some patches exist from this period to be worn by the enlisted personnel on the sleeve, however, it was not provided for in the decree, and the authors know of no valid examples of such a patch.

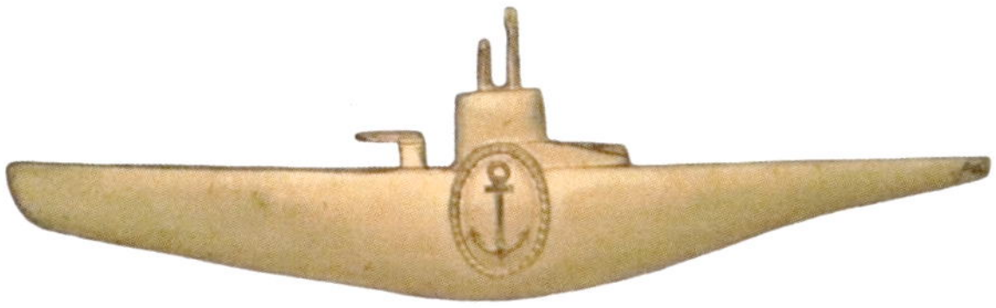

ARG-2.1

ARG-2.2

The officer's badge was modified in 1945 by replacing the Sun of May in the center oval with the national coat-of-arms, which is clasped hands in front of a pike with the Phrygian cap on top, against a field of the national colors. This is surrounded by a laurel wreath with a rising Sun of May sitting atop in stamped metal (ARG-3). This badge was the same measurements as the previous badges and also has a horizontal pin attachment. Miniature versions of this badge exist and are for use when miniature medals are worn and for berets (ARG-3m) It is not known by the authors when the mini was implemented. According to sources black cloth patches were issued for the officer badge around 1977 (ARG-3C.1 & 3C.2). Newer versions with bound boarders were issued beginning in the 1990s (ARG-3C.3).

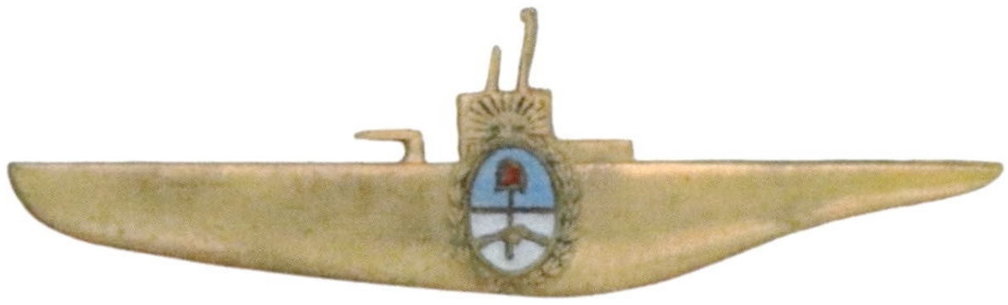

ARG-3

ARG-3m

ARG-3C.1

ARG-3C.2

ARG-3C.3

The enlisted badge remained the same design initially created in 1933 until 1969 when it was replaced with the silver version of the 1945 ARG-3 officer's badge (ARG-4) and mini (ARG-4m). At some point after 1970, first officers and then the ratings shifted to wearing the badge on the upper right chest. An original cloth version of this badge was issued with blue embroidery on white cloth (ARG-4C.1). Since then, other patch versions have been issued for use on a variety of uniforms in similar format to the officer patches (ARG-4C.2 & 4C.3). Sources identify that some patches are custom made and not standard issue, yet are still allowed to be used.

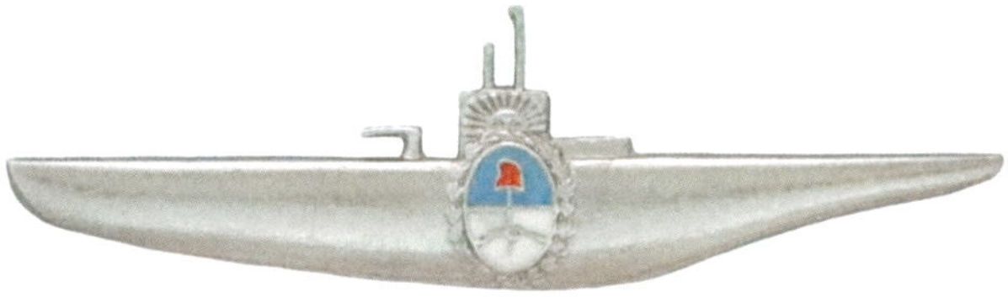

ARG-4

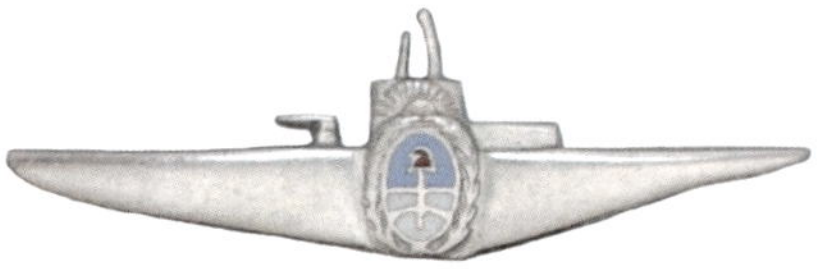

ARG-4m

ARG-4C.1

ARG-4C.2

ARG-4C.3

This status quo for metal badges was maintained until 10 December 1991 when a new badge was approved in the Navy Uniform Regulations, R.A.-6-001. This new badge was to be worn by medical and certain specialized engineering personnel identified as "Support Officers". This badge was gilt like the officer's badge, but the enamel national crest was replaced with a stamped metal version (ARG-5). In the 1990s, cloth versions of this badge were approved for use (ARG-5C.1 & 5C.2).

ARG-5

ARG-5C.1

ARG-5C.2

June of 1992 saw a new badge issued for the most Senior Chief Petty Officer of the submarine force. This was a horizontal rectangular pin with a miniature gilt officer's insignia on a field of black, with a gold rope border (ARG-6). This is worn below the qualification insignia on the upper right chest area.

ARG-6

There are silver and gold badges claiming to be for Argentinian submariners that have the same submarine profile, but instead of the symbols used for the officer and enlisted badges, a blue oval with an anchor is in the center. According to Argentinian sources, these badges have no official standing with the Argentine navy.

After the implementation of the Senior CPO badge in 1992, no further changes have been made to the Argentinian submarine insignias to date. Following the loss of the *San Juan* in 2017, Argentinian submarine operations were secured for the foreseeable future. The government is still pursuing options for new submarines and plenty of qualified submariners still serve on active duty, so the possibility still exists that we could see new badges issued sometime in the future.

AUSTRALIA (AUS)

The Royal Australian Navy (RAN) submarine force was officially established in 1914 with the acquisition of two British *"E"* class boats, the *AE-1* and *AE-2* that were only in service for a year. For the next 50 years, there were several attempts to reestablish an Australian submarine force, but most were short-lived due to various reasons. In the early 1960s the Australian government decided to fully commit to a submarine force due to the planned withdrawal of the 4th British Submarine flotilla from the country. Their first *Oberon*-class submarine was commissioned in 1967.

During the buildup to establishing the first Australian submarine squadron, the leadership of the submarine project team was convinced that submariners should have an insignia that was appealing, and of which they could be proud. Commander Alan McIntosh and Lt. Commander Henry Cook spearheaded the effort to create and approve a qualification insignia that would meet those criteria since the Australian submariners had seen the British "Sausage on a Stick" patch and were not impressed.

In 1965 they submitted a design for approval based on a concept developed by Commander McIntosh. This proposal did not receive much support until it reached the desk of Rear Admiral Victor Smith, who convinced the Naval Board to allow it to proceed. Following that approval, the team worked with Stokes of Melbourne to create the new insignia. A prototype was accepted by the Naval Board, and production of the new badge was approved.

This badge is a black shield with Her Majesty's royal crown atop, flanked by two inward-facing smooth, gold, bottle-nosed dolphins (AUS-1). On 25 July 1966, the RAN issued Navy Order #411 which addressed the eligibility requirements for wearing the qualification insignia, and the uniform regulations governing proper location. The badge is earned by a qualification process similar to those of most submarine forces. It is worn on the left breast of all uniforms by all personnel, regardless of rank, above the ribbon bar if applicable.

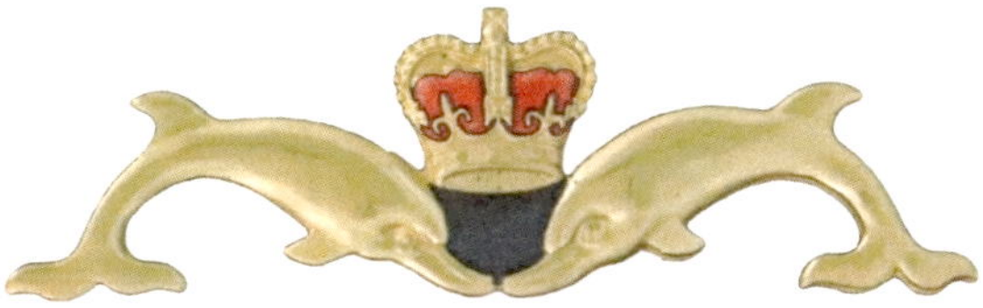

AUS-1

Only a small number of the badges were initially distributed. In 1968, the crew of the British submarine HMS *Trump* were given the badges to wear for a year. At the end of the year, they were asked to fill out a survey on the acceptability of the badges, and 99% of the crew approved of the design and wearing the badge. However, it was not until 1972 that the RAN officially issued the submarine qualification insignia to all qualified Australian submariners.

The front of this badge has remained completely unchanged since the original issue. Because there are several manufacturers there are some minor differences on the back of the badge, such as attachment method or pin placement, but the front design of all is identical. Initially the full-size badges had three clutch back pins (and the occasional horizontal pin). In the early 1980s, the three clutch back pins were phased out and only two were used, which is the current configuration. These badges are typically unmarked, however we will discuss the known versions with maker's marks. The original manufacturer Stokes has multiple versions of their mark, with the earliest saying just "STOKES" (AUS-1-STOKES-1.1 – 1.5). Another early manufacturer, G.A. Miller, has two known marks with one shown here (AUS-1-MILLER-1). A maker that no one has yet been able to positively identify, though "Christie's London" has been proposed, produced a rather poor-quality badge in the late 1970s (AUS-1-UK). The authors would appreciate a positive identification of this maker.

AUS-1-STOKES-1.1

AUS-1-STOKES-1.1r

AUS-1-STOKES-1.2r

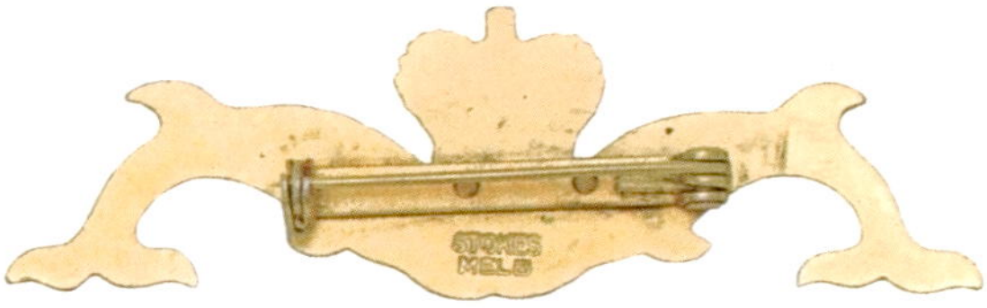

AUS-1-STOKES-1.3r

AUS-1-STOKES-1.4r

AUS-1-STOKES-1.5r

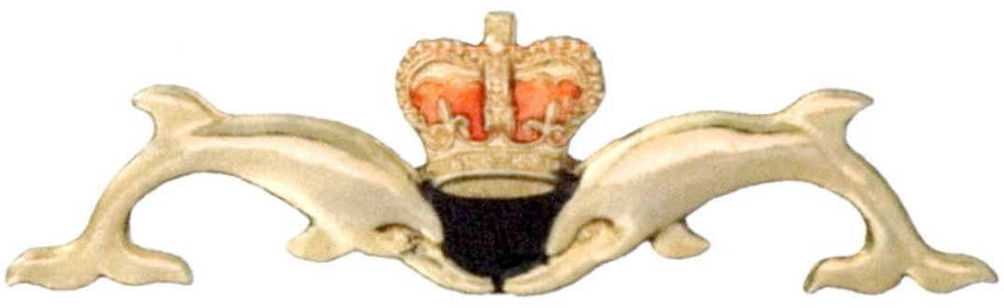

AUS-1-MILLER-1

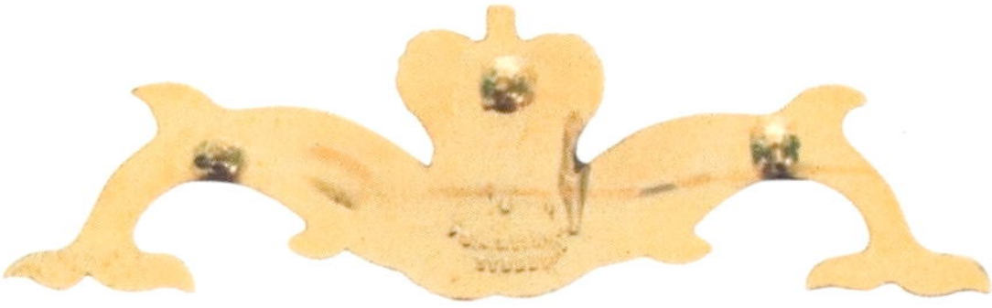

AUS-1-MILLER-1r

AUS-1-UK-1

AUS-1-UK-1r

There is a smaller, mid-sized version of this badge that is considered a mess-dress badge (AUS-1m). The known maker's marked versions of this badge are made by Swann & Hudson with two different maker's marks (AUS-1m-S&H-1.1 & 1.2), two different Stokes versions (AUS-1m-STO-1.1 & 1.2), and a "MILLER'S" marked badge (AUS-1m-MIL), but the vast majority are unmarked.

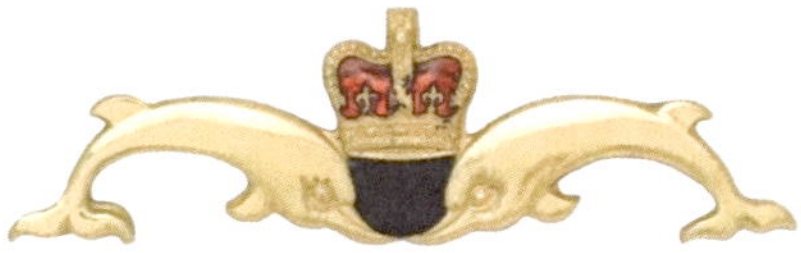

AUS-1m-1.1

AUS-1m-S&H-1.1r

AUS-1m-S&H-1.2r

AUS-1m-STO-1.1r

AUS-1m-STO-1.2r

AUS-1m-MIL-1r

On 19 August 1987, cloth patch versions of the insignia were approved for wear on work uniforms. The first approved was a lighter blue patch (AUS-1C.1), followed by a darker blue version (AUS-1C.2) and currently a black one (AUS-1C.3). Multiple minor variations of each of these patches exist. Bullion badges also exist in both full size and mess dress versions (AUS-VET-BUL & AUS-VET-BULm), but these are issued by veteran organizations and have no official standing within the military.

AUS-1C.1

AUS-1C.2

AUS-1C.3

AUS-VET-BUL

AUS-VET-BULm

Having what is considered by many to be one of the most simple and elegant submarine qualification insignia, Australia has kept their badge inventory relatively low by some standards, having only the full size and mess dress metal badges, and a patch for work uniforms. With the recent announcement that they will be adding nuclear powered submarines to its fleet, Australian submariners are training on U.S. vessels and earning U.S. dolphins… so, it is possible that Australia may be implementing additional badges in the future.

AUSTRIA-HUNGARY (AH)

Starting with the launching of their first submarines in 1908, the Austria-Hungarian submarine force only existed for 10 years, ending with the dissolution of the Austria-Hungarian empire at the close of WW1 in 1918.

The first submarine, or *U-boot*, badge for Austria-Hungary was implemented in 1910 as a ratings badge, issued only to enlisted personnel who were posted to submarines, making this one of the oldest verified submarine service badges. This badge belonged to the submarine, meaning, if you were reassigned or for some other reason left the submarine, you forfeited the badge. For this reason, and the fact that the badge was only in service for 8 years, a 1910-1918 period issued Austria-Hungarian badge can be very difficult to obtain.

AH-1

AH-1r

From its inception, the badge was a silver "UB" with the letters flanking and supported by a silver fouled anchor, with a wide clip or "spange" on the back used to attach it to the uniform on the left chest (AH-1). All badges issued from 1910 to about 1917 were standard, some of them with a "BSW" maker's mark, and some not.

Throughout the duration of their brief existence, these badges were die-struck. Near the end of the war, due to material shortages, some badges were cast of metal and used a vertical pin back as an attachment method (AH-1.1). There is evidence that some of these were actually made by the submarine crews themselves using sand casting methods as shown in AH-PHOTO-1. Verified versions of these theater-made badges are considered to be extremely rare.

AH-1.1

AH-PHOTO-1

Even though the Austria-Hungarian navy ceased to exist following the end of the war in 1918, there was a robust veterans' organization that organized multiple restrikes of the badge, some exactly the same as those issued from 1910-1918, including the hallmarks. Some had slight design and attachment variations, and some known and verified variants are shown (AH-VET-1.1 – 1.4). The practice of creating the veteran's badges continued for many years, and these badges were worn by ALL personnel who were assigned to submarines during their active period, including officers. Examples of these include jeweler-made versions some officers had made post-war (AH-VET-1.5).

AH-VET-1.1

AH-VET-1.1r

AH-VET-1.2

AH-VET-1.2r

AH-VET-1.3

AH-VET-1.3r

AH-VET-1.4

AH-VET-1.4r

AH-VET-1.5

AH-VET-1.5r

It is generally accepted that there are four sizes of the veteran badges. The original full-sized badge, and three smaller versions, with the smallest being a stick pin. These are all associated with the veterans' associations. (AH-VET-1 though VET-4). There are multiple other veteran badges that were issued, many with red and white enamel, but since these were not similar in appearance to the official badges, we have chosen to not address them here.

AH-VET-1-2, 3, & 4

During the late 1930s when Austria was working with Germany on rearmament, some officially sanctioned badges were issued to Austrian personnel serving on submarines in the Kriegsmarine so that they could be distinguished from their German counterparts. These badges are virtually impossible to differentiate from the originally issued badges, so provenance is key.

Because of the similarities between badges issued during these different eras, and with the addition of modern replicas, it can be difficult, if not impossible, to firmly ascribe a badge to a specific era without excellent provenance or the assistance of an expert. Original A-H and German submarine badges are some of the most copied badges out there, so, *Caveat Emptor*... "Let the buyer beware".

BANGLADESH (BAN)

In the period leading up to the official creation of the Bangladeshi submarine force with the transfer of two submarines from China to the Bangladeshi Navy in early 2017, some forward-thinking representatives in the newly created submarine squadron started the design process for a Bangladeshi submarine insignia. Led by Rear Admiral (Lower Half) Masud Iqbal, the team worked with several different manufacturers, including the Chinese shipyard making their submarines, to have an insignia made that reflected their vision.

These first prototypes were received by the squadron team in fairly small numbers. One known example likely made by the shipyard is a metal badge with a stylized black hand-painted submarine across a cast flat-faced silver anchor on a gold wreath with clutch back attachment (BAN-PTYP). Following this vetting process, the design was refined, and a manufacturer was selected.

BAN-PTYP

The first official Bangladeshi submarine qualification badge was issued to coincide with the receipt of the two submarines into the Bangladeshi Navy in early 2017. This badge is very similar to the BAN-PTYP but is more refined. The submarine has more intricate characteristics with three masts/periscopes and a rounded body of smooth gray metal on an angular polished silver anchor with a fully pierced anchor ring, the crown of the anchor nestled snuggly in the bottom of the much more defined gold wreath (BAN-1). Following a qualification process aboard the submarine, the Commanding Officer recommends the submarine candidate to Naval Headquarters for final certification as a qualified submariner. Once qualified, the recipient may wear the badge if they are assigned to other billets, unless they are removed from submarines for disciplinary reasons. The same badge is worn by all personnel, officers and enlisted, above the pocket on the left chest.

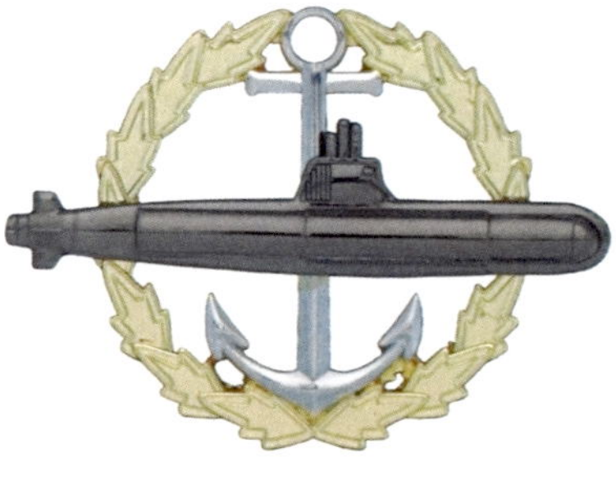

BAN-1

Around the end of 2022 a new Bangladesh submarine badge appeared. The new badge is a unique design in that it features a similar dark gray submarine and silver anchor from the previous badge, but the anchor ring is now filled with the national flower, *Shapla*, (a water lily) from the Bangladeshi national emblem. The anchor crown is still nestled in the gold wreath at the bottom, but the top arc of the wreath has been replaced with two gold leaping dolphins, their noses meeting at the *Shapla*. This badge is clutch back and, unfortunately, not official (BAN-UA-CP). Some submariners had it created as a "coat pin" that is worn by both submariners and non-submariners on their civilian clothing. With such a newly formed submarine service, it is possible that further Bangladeshi badge options may be explored in the future.

BAN-UA-CP

BRAZIL (BRZ)

Though Brazil ordered three *F-1* submarines from La Spezia, Italy as early as 1914, they did not have an "official" submarine force until it was created by Decree No. 19.747 on 10 March 1931. At that time, the uniform regulations were modified to sanction the wearing of a submarine badge.

From the beginning, both officer and enlisted Brazilian submariners have worn the same badge. This badge is modelled after the first *F-1* "unofficial" 1914 Brazilian submarines, showing the profile of a gilt, port-facing submarine riding on the surface through waves (BRZ-1). It is various shades of gold/brass depending on the manufacturer and age, and examples exist with the space between the antennae/periscope pierced (BRZ-1.1) and unpierced (BRZ-1.2). Some versions are much more refined than others (BRZ-1.3). It was worn, unsanctioned, by Brazilian submariners as early as the mid 1920s, and continues to be worn unchanged in basic design to this day, making it one of the longest surviving designs. There are also jeweler-made versions of this badge known to exist. It is worn on the left chest above the ribbon bar. Metal badges exist that are smaller versions of the BRZ-1; a stick pin and a "mess dress" size that are used for civilian wear but are not officially authorized for wear with uniforms.

BRZ-1.1

BRZ-1.2

BRZ-1.3

BRZ-UAC.1

Cloth patches were worn unsanctioned as early as the 1990s with examples known of gold on dark blue bullion patches worn on dress uniforms (BRZ-UAC.1) and gold thread on dark blue patches (BRZ-UAC.2) worn on working uniforms, with both versions sold at stores near the submarine base.

BRZ-UAC.2

Cloth patches achieved legitimacy in 2015 when they were authorized for wear on various work uniforms. Several variations were created for various uniforms as seen by the below examples (BRZ-1C.1 & 1C.2).

BRZ-1C.1

BRZ-1C.2

While it is unknown what the process for earning this badge was for the early Brazilian submariners, the current Brazilian submariner has an approximately 10-month long process to earn the badge with a "Qual card" that takes about 5 months to complete, followed by interviews, watch standing, and testing.

A badge featuring the same submarine profile, this time encircled by a serpent and with a diver's hardhat helmet in the center, is the official qualification badge of officers who complete the Brazilian Navy Submarine and Diving Medicine course (BRZ-2). Similar to the U.S. Submarine Medical Officer badge, recipients are not required to be qualified aboard submarines, but in submarine and diving medicine. Patches for this badge have been issued for various uniforms (BRZ-2C).

BRZ-2

BRZ-2C

On 26 October 2015, ordinance no. 89/DGPM approved the creation of the Brazilian Navy Submarine Psychologist badge. It is awarded to personnel who complete the Special Submarine Psychology Course for officers. This badge is the same gilt-colored port-facing submarine profile as the BRZ-1, but with a raised oval in the center, the Greek letter "PSY" inscribed on it (BRZ-3). Cloth versions were also issued (BRZ-3C.1 & 3C.2). To the authors' knowledge, this may be the only known Submarine Psychologist badge.

BRZ-3

BRZ-3C.1

BRZ-3C.2

Like other submarine forces, individual unit insignia exist, but only in patches so far with no known metal boat badges existing… yet.

BULGARIA (BLG)

Despite not having had many submarines of its own, Bulgaria has a fairly extensive and interesting submarine history. The first identified Bulgarian submariner was Lt. Rashko Serafimov who attended a Russian submarine navigation course from 1905-1907. At some point after his graduation, he was supposedly awarded a special badge. This badge, (RUS-SER) was donated to the Varna Naval Museum by his heirs in 1969 where it is displayed to this day. This badge is shown and discussed in the Imperial Russia/Soviet Union/Russian Republic chapter for obvious reasons.

The other graduates from the course taken by Lt. Serafimov were awarded the RUS-1 badge in 1909 when it was instituted by Imperial Decree. It is surmised that this badge was presented to Serafimov at a later time, and, due to its completely different design than the RUS-1, possibly created by a local manufacturer specifically for the lone Bulgarian to graduate from the Russian course. A few other examples virtually identical to this badge have been identified, so it is possible other Bulgarians serving in the Russian submarine force received similar badges.

The Bulgarians raised a flag on their first submarine around 1916, but no Bulgarian-specific submarine insignia was issued at the time. Bulgarian submariners did earn qualification insignia in WWI and WWII by serving in the navies of other countries. There are examples of Bulgarian officers earning the GER-1 badge of the Kaiser's Imperial German Navy following its issue in 1918 and into the 1930s as Bulgarian officers continued to attend submarine school in Germany. During WWII, while there is evidence to show Bulgarians served aboard submarines of both Germany and the USSR, there is no proof of any separate qualification insignia being awarded.

There are badges that have been credited with being worn by Bulgarian submariners during WWII (BLG-PROF-1.1 & 1.2) as evidenced by photos showing them, but there is no evidence that these were officially sanctioned, nor that they were anything other than just profile badges and possibly not even Bulgarian-specific.

BLG-PROF-1.1

BLG-PROF-1.2

It is worth mentioning that there are badges that have been erroneously identified as Bulgarian Submarine School badges. These are the badges that were issued to all graduates of the Bulgarian Naval Academy during a certain period in the late 1930s. Each badge has a submarine on it because Bulgaria was in the process of purchasing newer submarines during that period. There are at least two known badges, one with a "38" (BLG-NA1*) and the other with a "39" (BLG-NA2*), the latter one being the most copied/reproduced.

Following WWII, the first evidence of Bulgarians earning qualification insignia was for those officers serving aboard USSR submarines when they were issued the RUS-2 or RUS-3 badges. There were also examples of Bulgarian Commanders of Soviet submarines earning the RUS-4 badge.

While the Bulgarian submarine force was technically established in 1948 at a Soviet naval base in Odessa, they officially mark the start of their submarine service from 1954 when a submarine base was established at Varna. No official Bulgarian submarine qualification insignia existed until 1983, when a Captain in the submarine squadron had a badge designed. This badge is a 30mm x 40mm gold shield, a gold submarine with red star in the center on a sea of light blue below the green, red, and white horizontal stripes of the Bulgarian flag with another red star on the top left (BLG-1). He had 500 copies of the badge made at the Soviet mint, which is why on the back of this badge can be found "*Ц15k*". This was a typical mark for items made in Russia and represented the "C-price"… in this case, 15 kopeks. These badges have a safety pin catch. Collectors should note that badges with "*415k*" on the back are fakes.

After several years, the navy ran out of the 500 original copies, and a new badge was designed and manufactured in 1988. The new badge was a 30mm x 41mm gold shield once again with the white, green, and red stripes of the Bulgarian flag, but this time with a gold stylized submarine and a red star in the middle on a sea of darker blue (BLG-2). The fastening method for these are known to be either safety pin catch or screw back.

BLG-1

BLG-2

On the 45th anniversary of the Varna-based submarine squadron in 1999, a new submarine badge was issued. This is also a gold shield, 30mm x 38mm, but it has a gold admiralty anchor with a ship's wheel and a port-facing bow-on submarine in the middle, all on a dark blue enamel field (BLG-3). This badge has a screw back attachment method.

BLG-3

The latest version of the Bulgarian submarine qualification insignia was issued in 2008. This badge was similar in shape to the BLG-1 insignia, with a similar style submarine in the middle. However, there are several differences. First, the top white section of the first badge is replaced with light blue, and the light blue behind the submarine in the first badge is now replaced with rough seas in the same gilt color as the rest of the shield. Additionally, the gold of the submarine and the red star in the middle is now silver. Also, the letters, "*СПРБ*" have been added in the top light blue section, standing for "Union of Submariners in the Republic of Bulgaria". The size is slightly smaller at 29mm wide x 30mm high. Only 100 of these badges were issued (BLG-4).

BLG-4

Because so few of the 2008 badge were issued, the supply was quickly depleted, and in 2009 on the 55th anniversary of the submarine squadron, a second version of this badge was issued with two slight differences from the 2008 version. First, the light blue at the top was replaced with the white of the Bulgarian flag. Also, there was a "2" imprinted in the middle of the back of the badge (BLG-4.1*). This may well be the last Bulgarian badge issued as the Bulgarian submarine force was discontinued in 2011, and there are no current plans for adding submarines to the Bulgarian Naval register.

It is worth noting that Bulgaria also created multiple commemorative and veteran badges and medals. More information on these badges can be found in the paperback book, *Breast Decorations and Badges Awarded to Bulgarian Submarine Men,* by Petrov, Panayotov, & Zhechev, 2009, Sophia, BUL.

CANADA (CAN)

The Royal Canadian Navy (RCN) Submarine Service was established on 5 August 1914 with the purchase of two Holland class submarines that were originally built for the Chilean Navy. His Majesty's Canadian Ships (HMCSs) *CC-1* and *CC-2* were only in service for six years before being decommissioned in 1920, but they were the start of a long submarine history for Canada.

The RCN only commissioned four other submarines during the next four decades, and Canadian submariners also served aboard the British Royal Navy (RN) submarines during WWII. Prior to 1961, there was no Canadian submarine qualification insignia other than a cap tally denoting either the submarine's name, or "HM Submarine".

In 1955, near the beginning of the Cold War, the RCN entered into an agreement with the British Royal Navy to create a combination submarine squadron. The submarines were commanded by the Royal Navy, and half the crew was RCN. The Canadian crew members qualified just as the Royal Navy crew did and were certified to wear the current RN enlisted qualification insignia at the time, which was the cloth badge with the rather round submarine, also known as the "sausage on a stick" (UK-1).

In 1961, the RCN leased their first submarine from the United States and established their own stand-alone submarine service. With the christening of the submarine service came the desire for a Canadian submarine insignia. LCDR E.G. Gigg was the Commanding Officer of the first submarine. He worked with civilian navy artist Doug Baker to create "a distinctive Canadian submarine insignia clear of all the usual clutter of crowns, maple leafs, anchors, etc. already in abundance on our uniforms and which tend to obscure the designating symbol." Mr. Gigg wanted a "modern looking clean lined dolphin with its tail flippant in the air."

CAN-1.1

CAN-1.2

CAN-1.3

This first RCN qualification badge for submariners was a single heraldic dolphin of red thread, nose facing left and tail in the air, on a field of dark blue cloth (CAN-1.1). It was originally approved only for the enlisted men and was worn on the left sleeve of winter blues, and a royal blue dolphin on a white background was created for summer whites (CAN-1.2). Several months after being implemented, the officers were also allowed to wear the patch, a dolphin of gold bullion on a dark blue background (CAN-1.3), also on the sleeve above the Executive Curl.

LCDR Gigg initially had a gold metal version of the "Dancing Dolphin" or "Dancing Worm", as it became known, created as a sweetheart pin by Birks jewelers. This gold metal badge turned out to be perfectly suited for the officers to wear on their tropical whites, and it was adopted to fit that purpose (CAN-2) and was worn above the left coat pocket. These metal badges were almost exclusively manufactured by the Birks company, and multiple attachment variants exist, including broach pin, pin and lug, and clutch pins (CAN-2.1, 2.2, & 2.3), with the general consensus being that the clutch pin versions were made in the later 1960s. Full-size sweetheart versions of this badge were fabricated in silver metal by Birks with some iteration of "Birks Sterling" on the back and most known examples having single or double clutch pin attachment points (CAN-SW-1.1 & 1.2). Miniature versions of both were also worn as tie tacks or lapel pins on civilian clothing, but were not military issue.

CAN-2.1

CAN-2.1r

CAN-2.2r

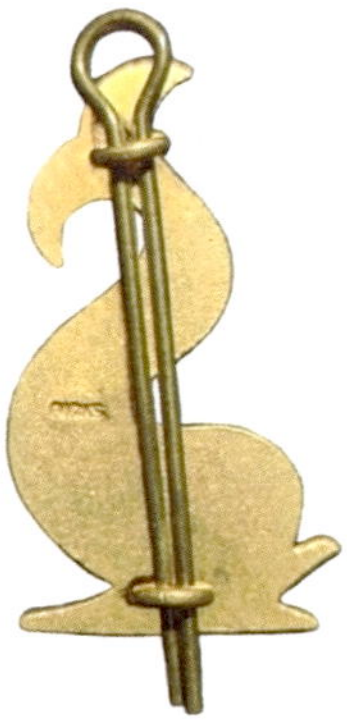
CAN-2.3r

CAN-SW-1.1

CAN-SW-1.1r

CAN-SW-1.2r

On 1 February 1968, the Canadian Armed Forces (CAF) was created. This institution absorbed all of the previously separate branches of the military, including the RCN, under a single authority. At this time, green uniforms were phased in replacing the blues, grays, and khakis.

The first CAF submarine badge was approved in 1970 and was issued in 1971. This was a single dolphin cloth badge with yellow thread on a dark green background (CAN-3.1). Many of these badges were such a dark green they appear to be black until compared to something black. From 1968-1971 the old-style single dolphin metal and cloth badges were still in use, and a gold bullion single dolphin on green background also appeared during this time. This gold bullion version was assigned an NSN (a NATO or National Stock Number given to all insignia authorized for wear), but formal approval documents for it have not yet been discovered (CAN-3.2). During this period, all versions of the single dolphin were moved from the sleeve to be worn on the left breast. Modern copies of these bullion badges have been made and the example shown here is one of them.

CAN-3.1

CAN-3.2

Almost immediately after the approval of the single dolphin badge for the CAF uniforms, the Commander of the Canadian Submarine Squadron decided, due to his dislike of the "yellow worm", that the submarine service needed a new badge. He had a design commissioned and it was quickly approved by naval authorities, and actually received royal approval when Queen Elizabeth signed an approval document in January 1972 (CAN-ATT1).

Approved Elizabeth R

Canadian Forces
Submariner Badge

Canadian Forces Headquarters
January 1972

Buckingham
Director of Ceremonial

CAN-ATT1

This new submarine qualification badge had been redesigned to include the relatively new maple leaf national symbol. This cloth badge was officially described as, "a crimson garnet wreath of laurel between two swimming dolphins in gold, above the wreath a crown, within the centre of the wreath a gold coloured maple leaf." This badge was on a dark green backing (CAN-4) and was approved for wear by all personnel, officers and enlisted. Initially, it was worn on the left sleeve, but, like the single dolphin before it, moved to being worn on the left chest of all uniforms, which is now the standard. Though officially approved in 1972, this badge was not widely distributed until 1975-1976. For the duration of its existence, this cloth badge remained basically unchanged, however, as a private purchase item some minor manufacturing differences can be seen to the crowns and dolphin shapes (see CAN-4.1 & 4.2).

CAN-4.1

CAN-4.2

The color of the backing was changed to black in 1985 (CAN-5) and this version is still in use today, also with multiple variants (CAN-5.1 through 5.3).

Mess dress versions of both cloth badges have been issued. The pre-1985 version (CAN-4m*) was typically on a dark blue or black backing and, strangely, very seldom on a dark green backing like the CAN-4 badges. The post-1985 versions (CAN-5m) are on a black backing. This can make it challenging to tell the difference between the mess dress versions from both eras when they are both on a black background.

CAN-5.1

CAN-5.2

CAN-5.3

CAN-5m

The first metal and enamel version of the double dolphin badge has green enamel surrounding the red/orange maple leaf (CAN-6) and was approved for use in 1976. There were multiple manufacturers of this badge, as shown by the sample hallmarks CAN-6.1 through 6.6. According to sources, the green was actually a mistake and was supposed to be black, however the badge retained the same design until 2006 when the green enamel surrounding the maple leaf was finally changed to black (CAN-9). The 2006 version remains the approved current metal badge, and also has multiple manufacturers.

CAN-6.1

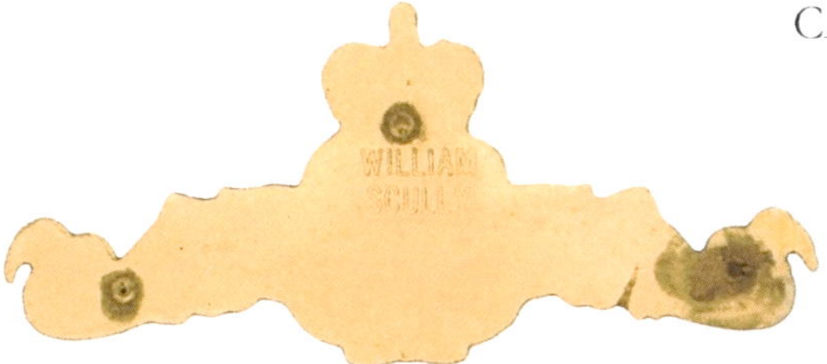

CAN-6.1r

CAN-6.2r

CAN-6.3r

CAN-6.4r

CAN-6.5r

CAN-6.6r

CAN-9

CAN-9r

There were miniature versions of the maple leaf metal badges created with both the green and black centers (CAN-6m & CAN-9m). These were originally approved for wear as a "Secondary Specialist Skill" badge in 1996. If a submarine-qualified individual subsequently qualified in another specialty, they would wear the badge for their primary assignment above their left pocket, and a miniature submarine qualification metal badge on the left chest pocket. Over time, these began to be used as mess dress badges as well, eventually completely supplanting the cloth mess dress badges in 2018.

CAN-6m

CAN-9m

In 1987 the CAF Army issued "Garrison Dress" uniforms. This uniform was only approved for use by the Army, but a decision was made by the Clothing and Dress Directorate that all of the CAF badges should be produced in the Garrison pattern. The Air Force and Navy skills qualification badges were never issued *en masse* because Air Force and Naval personnel were not authorized to wear the Garrison Dress uniform, however, because the Canadian Armed Forces allow "cross-over" trades, there are personnel who are qualified submariners who go on to serve in other branches while still authorized to wear their former qualification insignia. The Garrison pattern submarine qualification badge is "old gold", which is a gold thread version of the double dolphin badge on a dark green cloth background with a heavy dark green thread border (CAN-7). This "cross-over" situation also allowed creation of a cloth submarine qualification insignia on red background. This badge (CAN-8) is for wear on Army mess dress uniforms and is typically utilized by special forces personnel.

CAN-7

CAN-8

The Royal Canadian Sea Cadets is a civilian youth organization that is actually sponsored and administered by the Canadian Armed Forces and funded by the Department of National Defense. Though not unique, this Canadian organization may have been the only one of its kind to provide a submarine course and on the job training (OJT) aboard actual submarines. The cadets were given two patches to represent their accomplishments, both in a gray/silver thread on dark green gabardine. The first patch presented when a cadet successfully completed the submarine course was a starboard-facing submarine profile below a maple leaf (CAN-SC-1). The other patch presented upon completion of submarine OJT is styled after the Canadian submarine force insignia but is two inward-facing dolphins below a centered maple leaf, with the word "CADET" below (CAN-SC-2). These are the only official patches issued during the existence of the Royal Canadian Sea Cadets' submarine program, which was in place from about 1981 to 1991. A sketch of a badge similar to the CAN-SC-2 patch, but with a crown at the top, maple leaf in the center and "CADET" at the bottom is shown in the Sea Cadet Dress Regulations, however, this badge was never approved, nor produced. Metal badges or bright silver bullion patches that exist are not official and were created only for the collector's market.

CAN-SC-1

CAN-SC-2

CHILE (CHL)

Ordering their first submarines in 1910 from Electric Boat in the United States, the Chilean submarine force failed to launch due to the eventual cancellation of that order. However, kicking off their submarine force with an official reception held at the Boston Navy yards on 4 July 1917 for six *H*-class submarines received from the United States as part of a settlement with the British government, the Chilean Navy can boast one of the longest continuously operated submarine fleets in the world. These six submarines were received at a formal reception in Valparaiso on 20 July 1918.

In 1917, several months prior to the reception in the U.S., a Chilean delegation visited the Vickers-Armstrong shipyard in the United Kingdom to view a Royal Navy *H*-class submarine. To commemorate the event, members of the delegation received silver starboard-facing *H*-class submarine profile badges made by local jeweler, W.T. Story, Ltd. These badges were unmarked and in presentation boxes (CHL-H-PRES).

Later, after the submarines were delivered to Chile, unauthorized submarine qualification badges known as *"Piocha"* (insignia) appeared on Chilean submariners' uniforms. Sources identify that these "*Piocha H*" badges, as they became known, were a silver starboard-facing *H*-class submarine profile, complete with jumping wires, made by hand of carbon steel in the local torpedo workshop (CHL-1UA). It is also possible that jeweler-made examples of this badge exist due to the practice at the time of some personnel (mostly officers) having higher quality badges made. These badges were worn by both officers and enlisted personnel and were typically clutch-back. Due to the similarities between them, it is likely the W.T. Story Ltd. presentation pieces were the inspiration for the *Piocha H* badges, however there is no known documentation to confirm this. Modern reproductions exist in gold and silver.

CHL-H-PRES

CHL-1UA

After several years as an unauthorized accessory, the *Piocha H* achieved legitimacy with distribution of Supreme Decree No. 1659 on 25 August 1927. This decree, published in the Chilean *Navy Bulletin*, stated that the following requirement would be added to Regulation No. 25 for Chilean Navy Officer Uniforms:

> *"Article 51 a). Submarine specialist personnel will use a 60 mm long submarine as a badge of identification. This badge will be worn on the left side of the chest, above ribbons or medals and will be made of golden bronze for Officers and aluminum for Non-commissioned officers and crew".*

This decree now officially provided for a gilt-colored badge for officers (CHL-1*), and a silver badge for enlisted men (CHL-2*). These badges were essentially the same as the CHL-1UA, but officially authorized and made of different material. The authors have been unable to find any examples of these two badges, so it is possible that they were approved on paper, but never manufactured prior to the next generation of Chilean badges being issued.

Only a year after the *Piocha H* was officially authorized, the Chilean navy commissioned Vickers-Armstrong in the UK to build three new *O*-class submarines. Again, presentation pieces from a local jeweler, this time representing a silver port-facing *O*-class submarine, were given to a visiting Chilean delegation (CHL-O-PRES). Upon returning to Chile, this new badge quickly became the basis for a new Chilean badge. Because of the intentionally vague description of the Chilean submarine insignia in Article 51, and because of the pride the Chilean submariners took in their new modern submarines, the design of the badge was able to be changed to reflect the profile of an *O*-class submarine without modifying the Uniform Regs. While the *Piocha H* was still authorized for wear at this time, most submariners converted to wearing the "new" style badges.

CHL-O-PRES

CHL-O-PRESr

These *O*-class badges are a port-facing profile of the *O*-class submarine with jumping wires. These were also supposedly produced in gold (gilt bronze) for officers (CHL-3) and silver for enlisted personnel (CHL-4*). A Chilean jeweler-made badge is known in silver with a starboard-facing profile (CHL-4-STBD). While the Chilean uniform regulations called for badges in silver for the enlisted personnel and gold for officers as stated above, sources question if this was actually the practice or if there was a new regulation issued that supplanted the 1927 regs. This question arises because of the very small number of silver *O*-class badges known from that period, with significantly more gold badges; a statistic that doesn't make sense with enlisted personnel outnumbering the officers at least 4-to-1.

CHL-3

CHL-4-STBD

These badges were managed by the Supply Centers of the Chilean Navy and were provided to the submarine force crews from 1929 – 1971. There were a variety of manufacturers, and while most versions were cliché and used a horizontal pin attachment method, examples are known with clutch back attachment.

In 1971, a design change was implemented to the submarine insignias to eliminate the fragile jumping wires. This was done due to the deteriorated condition of the original die that had seen 42 years of continuous use. A parallel revision to the uniform regs approved (or reaffirmed) the gold badge for use by both officers and enlisted personnel (CHL-5). Following this change, several different versions of this badge were created by different vendors, using different materials, different finishes (polished or matte), and having solid or hollow back construction. Use of the jumping wires was even reinstated by certain manufacturers.

CHL-5

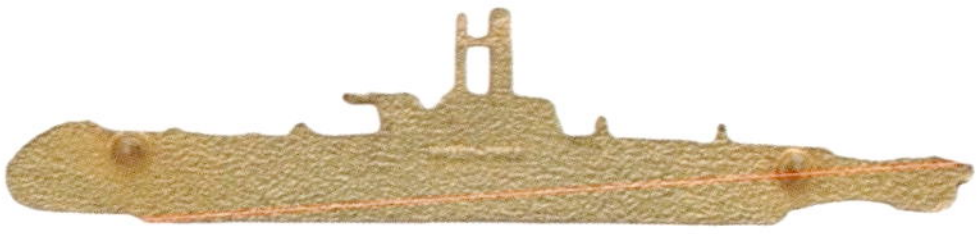

CHL-5r

In 1984 a silver badge of the same design was approved for use by Auxiliary Services such as Engineering, Medical, or Supply personnel who were not qualified on submarines, but were assigned to support them (CHL-6). It was only used by these personnel when they were assigned to submarine support billets and could not be worn if they moved on to another assignment. The use of this badge was discontinued in 1990.

CHL-6

CHL-6r

The current Chilean submarine insignia was approved in 1990. This was done in an attempt to standardize the badge, and to improve the quality of the metal, detail, and finish of the *O*-class submarine badge (CHL-7.1, 7.2, and 7.3) per Rule 7-31-67, Annex B of the Uniform Regs. This rule also formally reinstated the jumping wires. Since 1990, the manufacture of this badge has been entrusted to only a single manufacturer at a time, chosen by the Chilean Navy, and remains in use with no changes. It should be noted that, due to the similarities in design over the years, differentiating between the different time periods certain badges were used can be challenging without the assistance of an expert.

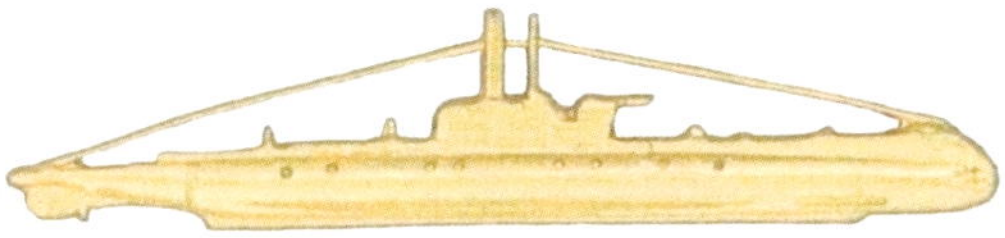

CHL-7.1

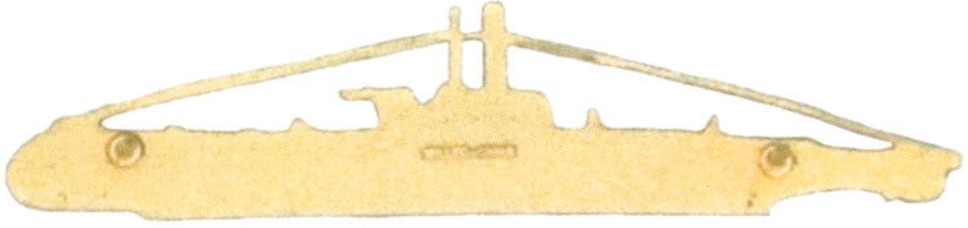

CHL-7.1r

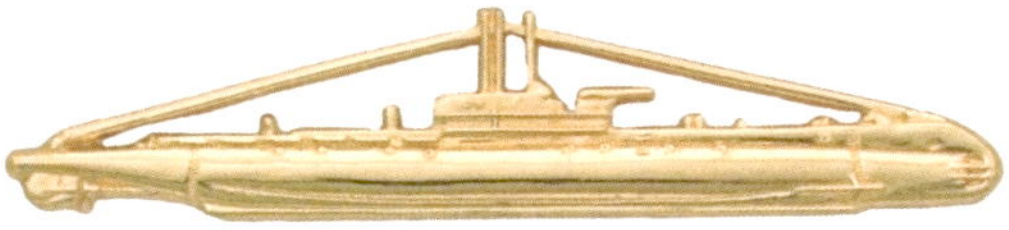

CHL-7.2

CHL-7.2r

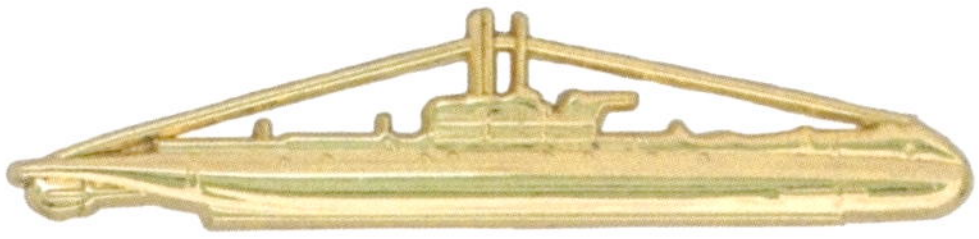

CHL-7.3

CHL-7.3r

There is also an approved submarine insignia for wear on the Chilean Navy uniform beret. This is a somewhat smaller *O*-class submarine insignia, surrounded by a wreath topped with the heraldic crown. Created in gilt metal in 1978 (CHL-8), this insignia is available in multiple versions. A variant in black has been created for the collector's market, but has no official use or standing.

CHL-8

CHL-8r

In the past several years, a cloth patch has been approved for use on submariners' working uniforms. This is a gold submarine profile on a dark blue background, and is the first patch authorized for use in any official capacity (CHL-7C).

CHL-7C

COLOMBIA (COL)

The Colombia National Navy created their submarine service 29 August 1972 by establishing the ARC (*Armada de la República de Colombia*) Submarine Flotilla to manage the logistics associated with multiple different submarines under contract. Two midget submarines from Italy were received in 1973, with one of those conducting the first dive of a Colombian submarine on 6 November of that year, now assigned as "Submarine Day" in Colombia. Two years later two Type *209* submarines were commissioned in the Kiel Germany shipyard 18 April 1975 and received in Cartagena shortly after. These submarines are still in service.

The Colombian submarine insignia was established in parallel with the submarine acquisitions. Based on an idea presented by CF Carlos Ospina, Admiral Manuel Avendaño Galvis, the first Commander of the Submarine Flotilla, designed the badge in late 1971 and it was approval by the National Naval Command in March 1972. This submarine qualification badge is a port-facing gilt Type *209 Pijao*-class submarine profile with the Colombian Navy coat of arms in the center on a silver fouled naval anchor supported by the noses of two gilt bottle-nosed dolphins (COL-1). The same badge is worn by all submariners, officer or enlisted, and is presented after successful completion of a two-year qualification process. It is worn on the left chest of all uniforms above the pocket or ribbon bar. The same badge is also presented to "Honorary Submariners", who, if military personnel, may wear the badge on the right chest of their uniform. Variations due to manufacturer exist (COL-1.1, 1.2 & 1.3).

COL-1.1

COL-1.2

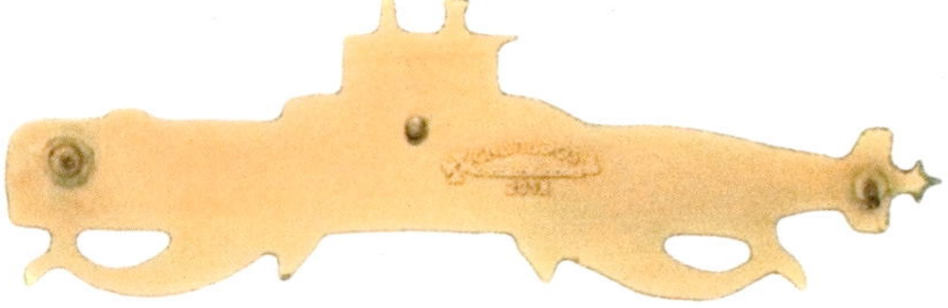

COL-1.2r

COL-1.3

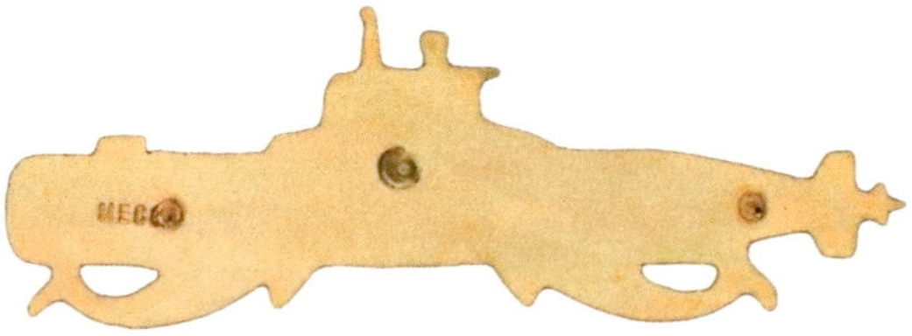

COL-1.3r

A miniature version of this badge is approved for use on mess dress uniforms, but the authors have been unable to identify when this was approved (COL-1m).

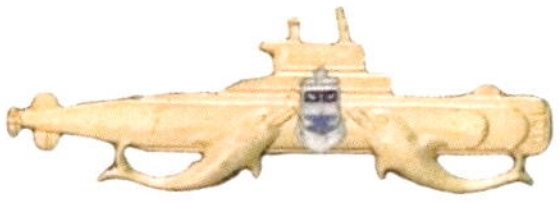

COL-1m

Colombian submariners that qualify on submarines belonging to other countries are authorized to wear the qualification badge of that country. This is most prevalent with German submarines.

There is a badge featuring a gold port-facing swimmer delivery vehicle with the same Colombian Navy coat of arms in the center. This is specifically for use by the Colombian commando personnel who operate their two swimmer delivery vehicles and is not associated with the submarine force (COL-COMMANDO).

COL-COMMANDO

Silver badges and black badges of the same design as the official submarine insignia are commonly found and are made by the same manufacturer who makes the official badges, however, according to official sources neither have any known official approval nor use within the Colombian National Navy.

A gold badge with an older port-facing fleet-style submarine profile with two inward-facing sharks under the hull and an enamel Colombian National Arms in the center has been identified by other sources as a possible early attempt at a Colombian submarine insignia. According to official sources, the COL-1 is the only official metal insignia that has even been used for the Colombian submarine service. While it is possible the badge with the sharks was an early unsanctioned attempt at a submarine insignia, the authors have found no documentation supporting this theory, and no validated examples of the badge. Copies of this badge do exist.

When camouflage working uniforms were implemented for submariners in the early 1990s, a patch featuring a black-outlined submarine profile on a green background was approved for use (COL-1C.1).

COL-1C.1

Newer patches feature the same submarine insignia with the word "ARMADA" under the insignia on a pixelated camouflage background (COL-1C.2) and also on a blue background with gold trim (COL-1C.3) for submarine coveralls.

COL-1C.2

COL-1C.3

Within the past 10 years, enlisted personnel qualified in submarines have been authorized to wear a submarine designation patch on their right shoulder in addition to the submarine qualification badge. All enlisted submariners wear a bright blue port-facing submarine profile on white patch on summer dress uniforms (COL-2.1) and the same profile in gold on a black background for winter dress uniforms (COL-2.2). For working uniforms, the first three levels of petty officers are provided with a patch featuring the same submarine profile in bright blue thread on gray material (COL-2.3), and senior petty officers wear the same patch with a blue submarine profile, but on khaki material (COL-2.4).

COL-2.1

COL-2.2

COL-2.3

COL-2.4

CUBA (CUB)

Though Cubans distinguished themselves in submarine service as early as the Spanish Civil War in the 1930s and serving aboard U.S. submarines during WWII, Cuba's first foray into developing a submarine force occurred in the late 1970s with the purchase/loan of three *Foxtrot*-class submarines from the Soviet Union. Though rumors persisted of them having submarines earlier, these could not be verified. Having had as many as six submarines through the 1980s, the Cuban submarine force currently has a single submarine – a North Korean midget submarine that has a questionable active status.

The Cuban submarine insignias were implemented in 1979 to align with the receipt of the three *Foxtrot*-class submarines from the USSR according to sources. A metal badge with the same port-facing submarine profile used for the Russian commander badge was issued in gold for Cuban submarine commanders, the difference being it had the Cuban coat of arms pressed into a shield mounted in the middle of the submarine instead of a red star (CUB-1). By the mid-1980s the practice changed to issue the gold badge to commanding officers and chief engineers, and a silver badge of the same design was created to be issued to the rest of the officers aboard (CUB-2). The coat of arms on these badges is simply pressed into the metal, and it is difficult to make out any details. Copies of this badge have been made, so finding a legitimate badge could prove to be challenging.

CUB-1

CUB-1r

CUB-2

CUB-2r

There are well-made reproductions that have a beautiful enamel blue and white Cuban coat of arms on the front. While high quality, these are reproductions, not authorized badges.

DENMARK (DEN)

The Danish submarine service existed without any qualification insignia from 1909-1971. Acquiring only their initial submarine from Italy, their following boats were constructed in Denmark. King Frederik the 9th, who, as a former Naval officer, was in favor of "clean" (insignia-free) uniforms, had opposed creation of any service-specific insignia. Following the King's death in 1972, a group of submariners led by LtCmdr P. B. Sorensen made unofficial embroidered submarine insignia badges and began to wear them. This badge was two gold outward-facing bottle-nosed dolphins, their tails supporting a seal with a fouled naval anchor in the center and a royal crown with red highlights. Only 60 examples of this badge were created (DEN-UA-1). A modern reproduction of this badge exists that is easy to discern because it does not have the red in the crown (DEN-UA-1R*).

DEN-UA-1

The Naval Command noticed this minor uniform regulations violation and, instead of issuing any disciplinary actions, they worked with members of the group to commission a version of the badge that was more in accordance with Danish heraldic tradition. An insignia was created with gold embroidery that was two bottle-nosed dolphins flanking a fouled naval anchor with a crown resting on the stock (DEN-1). It was approved as the Danish Submarine insignia in 1973 and is still the only officially approved insignia for dress uniforms to this day. To qualify for this insignia, a Danish submariner has to pass submarine training and qualify on the submarine. Serving aboard a submarine for 3 years or more allowed the recipient to continue to wear the insignia on the uniform even after leaving the submarine service and the badge is worn on the left chest above the ribbon bar if worn.

DEN-1

In 1988, a Spanish submarine insignia collector had a gilt metal badge created based on the Danish embroidered submarine insignia that he sent to the Danish Submarine Squadron Command. This badge prompted a lot of interest in having a Danish metal badge created. Taking quick but unsanctioned action, a small group of officers in the Submarine Squadron had the badge copied by a badge manufacturing company, and 99 samples created. These were received in 1989 and are referred to as the 1989 prototype badge (DEN-UA-2-PTYP).

However, the submarine community was not satisfied with the look nor quality of this prototype, and a craftsman was commissioned to refine the badge. This effort resulted in the first, yet still unofficial, metal submarine insignia being issued in 1990. This version was so widely accepted within the Submarine Squadron that, eventually, even the Squadron Commanding Officer wore the metal badge. Because the Spanish collector's badge had a hole just below the peak of the crown, the craftsman had mistakenly created a similar hole in the new badge. A batch of 100 of the first pattern of this 1990 badge were created. These were issued with clutch pin attachment points and a "waffled" or textured back (DEN-UA-2.1).

DEN-UA-2-PTYP

DEN-UA-2.1

Because the hole was considered a flaw, this was corrected on subsequent versions of the badge. The next version had the hole filled, but it was still noticeable from the front. This is the 1990 2nd pattern (DEN-UA-2.2), and, again, one hundred examples of this badge were created. The hole issue was completely rectified on the 3rd and final version, which is the current version (DEN-UA-2.3). Though never authorized, these badges were worn and tolerated for use by the leadership. A miniature version of the badge was created in the early 1990s, but has no officially sanctioned use and is used for wear as a lapel pin on civilian clothes.

DEN-UA-2.2

DEN-UA-2.3

In the late 1990s, versions for five (V) and ten (X) years of submarine service were created, but these never gained any popularity and were scarcely used (DEN-UA-3.1, DEN-UA-3.2).

DEN-UA-3.1

DEN-UA-3.2

The Danish submarine service was disbanded in 2004, however, former submariners continued to serve in the Armed Forces. In 2017 the Multi-Cam uniform was adopted for use as the standard work/utility uniform, and at the request of the submariners still serving, a fully sanctioned embroidered submarine insignia with Velcro attachment (or sewn option) was created for that uniform (DEN-2). The latest versions of this badge were made in camouflage or khaki, and are quite possibly the last official versions of the Danish submarine badge that will be fabricated (DEN-2.1 & DEN-2.2).

DEN-2.1

DEN-2.2

In 2023 a new Danish metal badge surfaced. The design is basically the same as the DEN-UA-2 badge, and is also gold, but with a marked difference. While the badge has the same dolphins flanking a fouled naval anchor, the crown sitting atop the anchor stock is a very finely detailed royal crown that is much more in the style that is appropriate for Danish Heraldic Tradition, and more similar to the official crown used on Danish national crests and logos than the crown used on the previous metal badges. As previously mentioned, Denmark has not had submarines since 2004, and according to military sources this is not an official badge. While it is possibly the work of a submarine veteran, direct contact with the Danish Submarine Veteran's Association determined that this is not a known initiative by anyone in that organization (DEN-UA-4).

DEN-UA-4

ECUADOR (ECU)

Ecuador has a history in submarine innovation and development dating back to 1838 with creation of the first successful submarine in Latin America, the *Hippopotamus*. But the Ecuadorian Submarine Force was not created until 139 years later when Ecuador purchased and commissioned two German Type *209/1300* diesel electric submarines between 1977 and 1978. These two submarines remain in service and are the only submarines the Ecuadorian Navy has ever had.

The Ecuadorian submarine qualification insignia was introduced in 1976 when the navy started the submarine purchasing process. The design of the first badge is a gilt port-facing Type *209* submarine profile with a single periscope protruding from the sail, in front of two gold oak boughs, with a silver condor from the Ecuador coat of arms in the center surrounded by a silver oval rope (ECU-1). The badge is worn on the right chest of all uniforms and has had various attachment methods including double screw back, clutch back and horizontal pin. The same badge was issued to all personnel regarding of rank.

ECU-1

In 1980, the badge was slightly altered when a snorkel mast was added to the top of the sail (ECU-2). Following two naval conflicts in 1981, the Ecuadorian Navy authorized the use of stars on the submarine insignia to indicate the number of conflicts the wearer participated in aboard a submarine. Initially there were two authorized conflicts in 1981, and one was added to the list in 1995, which allowed for three silver stars to be worn at that point (ECU-2.1, 2.2, & 2.3).

ECU-2

ECU-2.1

ECU-2.2

ECU-2.3

This changed in 2014 when the regulations were modified to allow one gold star in place of three silver stars (ECU-2.4). The badges are manufactured with the appropriate star combination so that submariners do not need to drill the badges to add the holes for attaching stars themselves. This badge remains the current Ecuadorian submarine insignia.

ECU-2.4

Miniature versions of the ECU-2 are authorized for use on the mess dress uniforms (ECU-2m, 2m.1, 2m.2 & 2m.3) and are also used by submarine officers wearing the dark blue side cap, or "piss cutter". Gold is the only color authorized for use in either full-size or mini. Though silver and black versions of these badges are abundant, they have no official standing and are not used by the military at all.

ECU-2m

ECU-2m.1

ECU-2m.2

ECU-2m.3

Submarine insignia patches are authorized for use on coveralls and jackets. Examples of several variations are known such as the dark olive submarine profile with a gold condor and wreath on a lighter olive background (ECU-2C.1), and gold submarine profiles on navy blue patches. The patches often have words such as "*Armada del Ecuador*" (Navy of Ecuador) in an arc above the gold submarine insignia, and "*Comando de Submarinos*" (Submarine Command) below. A patch has recently been added for the pixelated green camouflage work uniforms, which is the submarine badge profile in black with the condor and the oval in bright silver on the pixelated green camouflage background (ECU-2C.2).

ECU-2C.1

ECU-2C.2

In 2005 command badges were created for all of the different commands across the Ecuadorian navy. For the submarine force this is a small gold badge with the port-facing Type *209* submarine profile across a fouled naval anchor, surrounded by a thin circular gold rope. Initially, the commander of the submarine squadron was provided a badge that had either one (ECU-COM-1) or two (ECU-COM-2) stars on top, depending on their rank, and submarine commanders were given a badge with no stars (ECU-COM-0). The badge with no stars was discontinued in 2015 and submarine commanders were assigned the badge with the single star while squadron commanders now wear only the two-star badge. These are clutch back badges and are worn on the right chest of all uniforms, above other badges.

ECU-COM-0

ECU-COM-1

ECU-COM-2

EGYPT (EGY)

Though Egypt initially expressed interest in establishing a submarine force post-WWII, their submarine force did not officially begin until 1957 with the acquisition of seven *Whiskey*-class submarines from the U.S.S.R.

According to various sources, their submarine service badge was implemented as early as the late 1950s, or as late as the early 1960s, shortly after the commissioning of the *Whiskey*-class submarines. The first badge was a stylized angular port-facing gilt-colored submarine profile with an embossed Eagle of Saladin in the center from the Egyptian coat of arms and flag. This badge was eventually issued in four styles: The largest (80mm) for submarine Commanding Officers (EGY-1), a medium size (55mm) for submarine officers (EGY-2), a smaller one (45mm) mounted on a round ring for petty officers (EGY-3*), and the same size but without the ring for ratings (EGY-4*). Because of the lack of valid examples, it is possible the EGY-3 & 4 badges were discontinued well before the EGY-1 & EGY-2 badges. These badges were all stamped and hollow back with horizontal pin fasteners. They were worn on the upper left chest of the uniforms, above the medal bar when present.

EGY-1

EGY-2

In 2011 Egypt purchased two *209*-class submarines from Germany. This appears to have been the impetus for a badge change with new badges issued that year. A badge was authorized that, similar to the previous badges, featured a port-facing submarine profile with an Eagle of Saladin in the center below the sail – some have been seen with a silver eagle, and others with the eagle just as a gold embossment. But that is where the similarities end. This new gilt-colored badge is a more detailed *Romeo*-class submarine profile, prominently featuring a bulbous bow, handrails along the deck of the submarine, and a stepped sail. The same profile was issued in three version: An 80mm version issued for submarine Commanding Officers (EGY-6), a 50mm version for officers (EGY-7*), and an even smaller 30mm-long profile lacking many of the details of the larger badges due to its size is said to be issued to petty officers (EGY-8*). These are all clutch back attachment badges and are worn in a manner similar to the previous badges.

EGY-6

There are photographs of a basic port-facing submarine profile patch used by Egyptian submariners, located on the left chest above the name tag on working uniforms. The authors have been unable to find a verified physical example of this patch (EGY-9C*). To the authors' knowledge, these are the latest known Egyptian submarine insignias.

For years there has been a badge sellers purport to be an Egyptian submarine badge. This stylized gilt (and silver examples are known) port-facing submarine profile has multiple ovals and some Arabic numerals on the side of the submarine (EGY-FAKE). A version is also known that is attached to a metal circle similar to the EGY-3. Per official sources, this badge is fake, having no affiliation with the Egyptian Navy, and is often found being sold in the bazaars of Cairo and Alexandria. These are hollow back with horizontal pin attachment. While not many fake badges are shown in this book, this one is so widely accepted as valid we felt it important to share just for the purposes of debunking it.

EGY-FAKE

There does not appear to be any official regulations governing the issuance or appearance of the Egyptian submarine badges, so it is possible that new badges can be created if the current version falls out of favor.

FRANCE (FRA)

France was one of the early innovators and adopters of submarine technology, building their first submarine in 1863. They then went on to being the first to adopt Diesel engines in 1904 for their submarine Z, amassing the largest fleet of submarines in the world in the first decade of the 20th century, and also building submarines for numerous other countries. Because of their prominent place in submarine history, it is fitting that the French Submarine Force issued the first verified submarine distinctive in the world.

In a French Navy official bulletin issued on 21 February 1906, a full three years prior to the RUS-1 badge, a ministerial decree was published associated with recruitment and organization of enlisted personnel. Article 14 of this decree spelled out the details for a special cloth badge that would be worn by enlisted personnel serving aboard submarines:

"The personnel embarked on the submarines carry on the left arm a special insignia representing a torpedo and two thunderbolts. This insignia is embroidered in gold for the petty officers, in red silk for the quartermasters and sailors".

The original decree was reinforced when the badges were mentioned in another Naval bulletin four months later on 26 June of the same year with a slight description change saying the badges would, *"consist of two torpedoes crossed with lightning"*. As described in the decree, there were two badges. The first, issued to all enlisted personnel who were not petty officers, was a design of crossed torpedoes with lightning bolts emanating from the center, embroidered in red silk (FRA-1.1). The second badge, for petty officers, was the same design, but embroidered in gold (FRA-2.1). Versions of both with sequins were also used (FRA-1.2 & FRA-2.2). These badges were to be sewn to the upper left sleeve of the uniform and had to be purchased by the submariners.

FRA-1.1

FRA-2.1

FRA-1.2

FRA-2.2

Because of the relatively vague wording of the badge design standards, multiple versions of these badges exist with different positions and shapes for the torpedoes, and varying numbers, colors, and positions of lightning bolts. Coinciding with the creation of "specialty" or ratings badges in the French Navy, versions of these badges began to appear with the letters, "M", "E", or "T" in the middle of the crossed torpedoes. These letters signified "Mechanic", "Electrician", and "Torpedoman" respectively, but were not consistently used by all submariners (FRA-1-E is shown). The date these were first implemented is not clear, but photographs showing them in use as early as the mid-1920s exist.

FRA-1-E

These cloth badges were in use until 1946 when, in typical government fashion, guidance was issued that caused confusion in the ranks. In June 1946 uniform guidance was issued requiring "specialty" or ratings badges to be sewn on the upper left arm. This raised questions from the submariners concerned that the ratings badges would replace their submarine specialty badge. A clarification was issued two months later in August 1946 changing the name of the submarine specialty badge to the *certificat d'aptitude à la navigation sous-marine* ("badge of the certificate of aptitude for underwater navigation") or CANSM, but this did nothing for the confusion associated with the badge location. It was finally resolved in December 1946 when a new metal badge was created, replacing the FRA-1 and FRA-2 badges.

The new metal badge for holders of the CANSM was a silver compass rose with a gold port-facing submarine profile with a step cut out of the forward section of the conning tower lying horizontally across the center, and a vertical silver sword with a gold hilt, tip up, behind the submarine (FRA-3). The attachment method was most generally a vertical pin mounted behind the sword, but horizontal pins are also known. This was to be worn by all personnel serving aboard the submarines, including officers, on the right chest of the uniform. These badges were numbered and assigned to specific individuals. The numbering continued until approximately 1968, though the requirement to do so seemed to vanish in 1957. Badges 1-2000 carried the assigned number on the back top left of the badge (FRA-3.1). Badges 2001 and higher have the number on the back lower right (FRA-3.2).

FRA-3.1

FRA-3.1r

FRA-3.2

FRA-3.2r

The FRA-3 was the only badge issued to French submariners until 1961. In guidance issued 18 June of that year, an "Elementary Submariner" badge was created. This badge had a smaller, more stylized silver compass rose with the same port-facing gold submarine profile with a step out of the forward corner of the conning tower (FRA-4). When first issued, it was worn only by enlisted personnel until they qualified on the submarine. There are fake versions of this badge with the submarine mounted on a round silver ring. The FRA-3 remained in use but was now called the "Superior Submariner" certificate.

FRA-4

FRA-4r

In December 1974, an overhaul of the French badges was approved and implemented in 1975. The Elementary Submariner badge was modified to have the full-size compass rose featured on the FRA-3 and a gold port-facing submarine profile, the step now removed from the sail to represent a more modern submarine (FRA-5.1, 5.2, & 5.3). This was now worn by enlisted personnel holding a Basic Submariner certificate, and officers with a certificate of General Submarine Knowledge.

FRA-5.1

FRA-5.1r

FRA-5.2r

FRA-5.3r

The Superior Submariner badge remained basically the same as the FRA-3, however the step was also removed from the sail making it easy to spot post-1975 Superior Submariner badges (FRA-6.1). This badge was now worn by enlisted personnel holding the Superior Submariner certificate and by officers holding the certificate of aptitude for underwater navigation and was made by multiple manufacturers. Beginning in 2008 a "renumbering program" was implemented that allowed personnel to have serial numbers etched on the back of the vertical sword blade of these badges (FRA-6.2). This numbering identified the year, the base where awarded (Toulon or Brest) and then the sequential number. It was allowed to be retroactive on all of these Senior Submariner certificates going back to 1968.

FRA-6.1

FRA-6.1r

FRA-6.2

FRA-6.2r

During the 1974 overhaul a new badge was created for submarine commanders. Following the same basic design as the FRA-6, this badge differed by having two crossed silver swords with gold hilts behind the submarine (FRA-7). The badges issued to submarine COs were serialized per the instruction with a number on a small flat spot on the lower right back of the badge (FRA-7.1r & 7.2r). This sequential numbering for the CO badges has continued to the current time. The COs can purchase additional badges that are not serialized (FRA-7.3r & 7.4r). At the time of publication, there have been no further changes to French submariner badges.

FRA-7.1

FRA-7.1r

FRA-7.2r

FRA-7.3r

FRA-7.4r

Though multiple cloth patches/badges have been used by French submariners for the past 80 years, there are none used by French submariners in an official capacity that can be used as a *certificat d'aptitude à la navigation sous-marine* badge substitute.

Beginning in the late 1930s, French submarines began creating boat badges. Many of these are small works of art, and some are quite rare. While we will not address them here, a listing of these pins/badges can be found in Jean-Pierre Stella's comprehensive guide, *Guide de Insignes des Batiments de la Marine Nationale de 1936 a 1970, 11e fasicule,* Paris, 1989.

GERMANY (GER)

Though having an early foray into submarine technology with a failed submarine attempt in 1851, Germany was a relative latecomer to modern submarines among major naval powers at the turn of the century with its first submarine, the *U-1*, not being commissioned until 1906. But the Imperial German Navy more than made up for its late start with its production capabilities, having 28 *Unterseeboote* or *U-boote*, some of the most technologically sophisticated submarines in the world at the time, operational or under construction in 1914 at the start of WWI. By the end of the war, Germany had put 346 submarines into service.

It was during WWI that Kaiser Wilhelm II approved the first Imperial German submarine "war badge" on 1 February 1918. Designed by Walter Schott, this gilt-colored bronze badge was a port-facing U-boat profile mounted across the middle of an oval wreath of laurel leaves wrapped in ribbon and topped by the German State crown, or *Reichskrone* (GER-1). Because the badge was cast in bronze with gold plating, the laurel wreath often formed a green patina that the wearer needed to polish away. Initially this badge was awarded to submariners regardless of rank completing at least three war patrols with exemplary performance, but that requirement was later reduced to two war patrols and 150 days at sea, or being wounded or killed in action. It was worn on or below the left chest pocket of the uniform. It was only officially in use until November 1918 and the official end of WWI, however, many badges were issued to veterans post-war. It is believed about 5,000 of these badges were awarded. This badge was also worn by some WWI veterans who ended up serving in WWII. Some hollow back versions of this badge were issued post-war to veterans

GER-1

As Nazi Germany began scaling up military capabilities in the 1930s, submarines were naturally included due to the successes they had in WWI. In the late 1930s, Germany just had 57 submarines in service. By the end of the war, they had built 1,190 U-boats. On 13 October 1939 a new German U-boat war badge was issued for Kriegsmarine submariners. The new war badge was designed by Paul Casberg of Berlin, and paid homage to the first German submarine badge. This gilt-colored badge had a starboard-facing U-boat profile mounted on a laurel wreath, but a newer style U-boat, and the wreath is only wrapped with ribbon in an "X" at the bottom. At the top of the badge was a German imperial eagle sitting astride the Nazi swastika (GER-2). Early versions of this badge are made of bronze, with versions from later in the war being made of a "gold washed" zinc. The badge was issued to submariners conducting two war patrols and was worn on or below the left chest pocket. As a patrol badge it had certain exceptions for being awarded, including being wounded, as an award for bravery, or posthumously if lost at sea. It is believed around 9,500 of these badges were issued.

GER-2

Embroidered and bullion fabric versions of the WWII German submarine war badge were also worn by the submariners, in the same place on the left chest, but typically more for work uniforms (GER-2C.1 & GER-2C.2).

GER-2C.1

GER-2C.2

The supreme commander of the German U-boat fleet, Admiral Karl Dönitz, had a special version of the GER-2 badge presented to him. This badge was the same design but was solid gold and had diamonds inlaid in the swastika and in the wreath. Following the war, the Dönitz badge was broken up and sold to pay for his legal fees. Following the receipt of his badge, Dönitz had special examples made that were presented to recipients of the Knights Cross of the Iron Cross with Oak Leaves and above. Like his badge, this special version was the same design as the basic GER-2 badge, but was made of gold-plated silver, with nine small diamonds in the swastika (GER-2-29). This badge was manufactured by Schwerin and also Gebr. Godet & Co. and is believed to have been awarded 29 times.

GER-2-29

On 15 May 1944, Admiral Dönitz had a new badge created to recognize Kriegsmarine submariners for at-sea duration. A bronze "combat clasp" was presented to personnel having participated in at least 15 war patrols and with 90 days at sea. This was a smaller version of the GER-2 badge with a fan of oak leaves extruding from each end (GER-3). Because the submarine force was spending so much time at sea near the end of the war, a few months later on 24 November 1944 a silver version of the same badge was issued to recognize submariners exceeding the bronze badge requirements and having participated in at least 25 war patrols and with at least 120 days at sea (GER-4). A gold version was planned, but never created nor awarded, and existing gold versions are fakes.

GER-3

GER-4

Also in 1944, the Shipyard Achievement pin was implemented. Not a submarine badge, it was only awarded to shipyard workers in France (La Palice, St. Nazaire, Lorient, and Brest) for special achievements in the repairs of damaged submarines. It is a stick pin with a submarine conducting a starboard-facing transient through a gear wheel, the German Imperial eagle at the top of the gear. It was issued in two versions, brass with the submarine profile showing 13 limber holes (GER-SY-BR), and aluminum with the profile showing 20 limber holes (GER-SY-AL*).

GER-SY-BR

At this point, it is of vital importance for collectors to be aware that <u>*all*</u> of the German badges previously discussed are among some of the most copied/duplicated/counterfeited submarine badges. While good amounts of each of these badges were issued (most in the thousands), easily tens of thousands of duplicates exist. Provenance is paramount when searching for genuine examples of these badges.

Note that, like Italy, the WWII Germany Navy had multiple other badges associated with commandos and mini submarines. There are other excellent sources of information for these insignia, and we will not address them here.

Following the war, Germany was banned from having submarines. It was not until 1955 that the Federal Republic of Germany's (West Germany) *Bundesmarine* received approval to develop submarines for coastal defense. The first two submarines they operated were U-boats previously sunken during the war that were raised and overhauled.

On 26 July 1957, a law was passed allowing West German WWII veterans to wear their awards in a de-Nazified form as the swastika was (and remains) a forbidden symbol. Many wore the WWII war badge or the combat clasp leaving the eagles, but with the swastikas either physically removed, or completely defaced creating new badges (GER-5, GER-6*, & GER-7).

GER-5.1

GER-5.2

GER-7

De-Nazified badges were created for all three that raised the U-boat higher within the wreath and completely removed the eagle and swastika (GER-8, GER-9, & GER-10).

GER-8

GER-9

GER-10

A badge specifically crafted for the U-Boat Veterans Association (*Verband Deutscher U-Bootfahrer* or VDU) in 1957 was a gilt WWII war badge, but with an enameled Iron Cross replacing the eagle and swastika (GER-VET-57). While not copied to the extent of the earlier listed badges, many of these badges have also been duplicated.

GER-VET-57

Beginning in 1960, young German submarine officers were sent to other countries, namely France, the United States, Great Britain, and Turkey, to train aboard submarines. These officers were granted permission to wear the qualification badges they earned from the associated countries.

It was not until 1966 that West Germany approved their first submarine badge, which was a trades badge (*Laufbahnabzeichen*) worn on the upper sleeve by lower-ranking enlisted personnel and on the lower sleeve by petty officers. This was issued to enlisted submariners upon completion of basic submarine school and was typically a round patch (though rectangular versions are known) with a starboard-facing stylized submarine profile in the center, atop two stylized waves. It was issued in three versions: red on dark blue (GER-11.1), red on white (GER-11.2), and gold on white (GER-11.3), the red being used for working uniforms, and the gold being used on the dress uniforms.

GER-11.1

GER-11.2

GER-11.3

On 7 April 1972, a new submarine qualification badge was issued for all ranks. It was a stylized port-facing *205*-class submarine profile laying in front of a laurel wreath, open at the top. The badge was created only as a patch and was initially only available in machine-embroidered versions, however hand-embroidered and bullion versions quickly became available. These were gold on dark blue for winter dress blues (GER-12.1), and gold on white for summer whites/dress whites (GER-12.2). They were worn on the right chest of the uniforms and were awarded to personnel after being attached to a submarine for at least six months. Because there could be a significant period of time between basic submarine school and actually earning their submarine qualification badge, many enlisted personnel chose to continue wearing their *Laufbahnabzeichen* trades badges until fully qualified. This practice continued until about 1981.

GER-12.1

GER-12.2

It was not long after the authorization of these new cloth badges that unauthorized metal versions began appearing. These bootleg gilt metal badges were various sizes and attachment points due to the multiple means of manufacture (including self-made), but typically stayed true to the approved design (GER-12-METAL-UA-1.1 through 1.4). Despite the use of these badges, it was not until 1984 that a metal submarine qualification insignia was approved for use.

GER-12-METAL-UA-1.1

GER-12-METAL-UA-1.2

GER-12-METAL-UA-1.3

GER-12-METAL-UA-1.4

This stamped badge is similar to the cloth badge in that it has the port-facing submarine profile lying across an open-topped laurel wreath, but instead of being like the unauthorized metal badges, this design was in relief on a plate with a pebbled finish (GER-13) with a clutch pin attachment. To say that this badge was not well-received by the German submarine force is an understatement – it appears to be almost universally disliked and is ignored to a large extent. However, despite its lack of popularity, it still remains the only officially approved German metal submarine insignia because it was issued at the same time as multiple other specialty insignias in the same style. It can also be issued as an honorary badge for individuals meeting certain criteria.

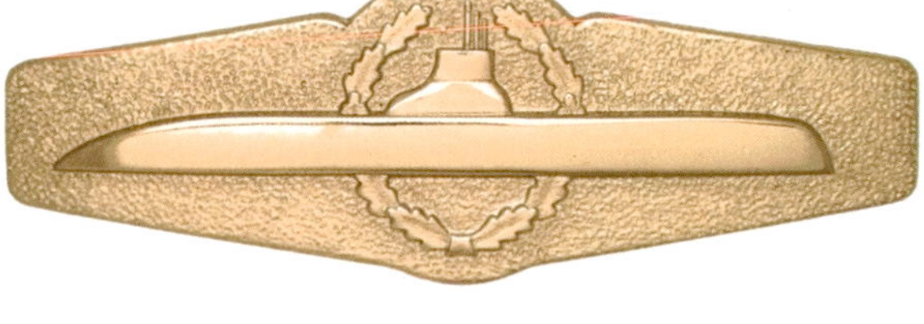

GER-13

In 1990 the Commander's star was introduced to the German Navy. An effort to introduce a submarine commander's badge incorporating the star and the submarine as a single badge the next year failed, so any badges with this combination are unauthorized fakes. With the introduction of the olive drab combat rig in the 1990s, olive drab and camo versions of the GER-12 badge were authorized for use (GER-12C.1 & 12C.2).

GER-12C.1

GER-12C.2

Despite wide-spread use in the submarine force since the 1970s, the metal version of the GER-12 has never achieved status as an officially approved badge… however, it may have just edged closer to that goal. In October of 2023 the Commander of the German Submarine Flotilla gave written permission to the submarine force to wear the latest version of the metal *212A* badge, and the submarine school is keeping a list of numbered qualification insignias presented. While this is just shy of official authorization, it is better than just the tacit acknowledgement that previously existed. This has started an effort by German submariners and veterans to have the badge officially approved, so we will preemptively grant this badge official status as GER-14. If this effort fails, we will rescind that status and make a retraction in the next edition. At the same time the GER-14 began to be issued, a corresponding patch was authorized for use. Having the same design and submarine profile, but in gold thread, this patch is worn on work and combat uniforms (GER-14C).

GER-14

GER-14C

As with several other countries, Germany has had some beautiful boat pins that have been issued to crew members. Because there are so many of these, and multiple variations for each submarine, we will not cover them here, and hopefully a comprehensive resource will be generated at some point. As with the other WWI and WWII badges, the boat pins for any U-boats operating during the wars are highly copied.

GREECE (GRK)

The Greek Navy may have a legitimate claim to the world's earliest documented use of submerged chambers with the army of Alexander the Great making military use of them as early as 322 BC. They were also on the leading edge of relatively modern submarine warfare, purchasing the first *Nordenfeldt*-class submarine in 1885 to create an advantage over their ancient foes in the Ottoman Empire, but there is no evidence that this submarine was ever commissioned by the Greek Navy. It was not until two submarines were purchased from France in 1910 and commissioned in 1912 that the core of the Hellenic Navy Submarine Command was formed. Shortly after their purchase, the HS *Delphin* became the first submarine to launch a self-propelled torpedo attack on an enemy ship during the first Balkan War.

The implementation history of the Greek submarine badge is much more difficult to ascertain than their submarine history, and there is disagreement even within the Hellenic Naval historian community, but there are a few things that are known for certain. Royal Decree #185 issued 15 June 1912 made allowances for the first Greek submarine insignia. Shown in the decree was a port-facing submarine profile, however, since it was required to be worn on the left sleeve of all uniforms below the rank insignia, it is likely that starboard-facing profile badges were created, in gold embroidery for non-commissioned officers (GRK-1) and in red thread for ratings personnel (GRK-2) as shown in these sketches. It is not known when metal badges began to be used, nor were authorized, but sources place the use of handmade metal badges as early as 1916, typically worn on the upper left chest, and of inconsistent design.

GRK-1

GRK-2

In the mid-1920s the Hellenic Navy began a concerted effort to expand their submarine force purchasing two *Katsonis*-class submarines from the French in 1925 that were commissioned in 1927 and 1928. Those were followed up with the purchase of four *Proteus*-class submarines, also from the French, that were commissioned from 1929 through 1930. These were the only six submarines operated by the Greeks until near the end of WWII when they received six British submarines between 1943-1946.

Sources identify that the Greek submarine badges may have become slightly more consistent in their design during the 1930s, with submariners wearing profile badges representative of the submarines on which they served. These same sources claim that the badges were made by hand from 1936 until the early 1950s by a petty officer on the submarine base, which may account for the multiple small design deviations between badges from this period, with no two exactly the same. Hellenic Navy Uniform Regulations from the mid-1930s provided for metal badges without a photo, but with a vague description. First, a gilt officer's badge, a submarine profile in "semi-submersion" that is 65mm long. The regulations do not provide additional details other than it is to be worn on the left chest above the ribbons. An NCO badge was provided for, also gilt metal, but 50mm long. And, finally, a badge for other enlisted personnel, this time a silver metal badge 50mm long.

A *Katsonis*-class profile badge shown in the 1939 Hellenic Navy Uniform Regulations (GRK-3.1-PHOTO) is believed to represent the officer badge of personnel serving aboard those submarines, however, no examples have been found by the authors other than this photo.

GRK-3.1-PHOTO

Similarly, while NCO (GRK-4.1*) and enlisted (GRK-5.1*) versions may exist the authors have found no physical examples. *Proteus*-class profile badge examples are known in at least officer (GRK-3.2) and enlisted (GRK-5.2) versions, and are identified in multiple photos of 1930s – 1940s era Greek submariners. While no such examples have been discovered by the authors, it is assumed NCO versions (GRK-4.2*) exist due to their presence in the uniform regulations.

GRK-3.2

GRK-5.2

More commonly, profiles badges are known for the British *V*-class submarines provided to the Hellenic Navy beginning with the *Pipinos* in 1943. The examples shown here are from the *Pipinos* (GRK-3.3.1), though with a cut off periscope and deck gun, and from a different *V*-class boat, probably around 1945 (GRK-3.3.2). These are different from the previous two styles in that the gun was moved from the sail to the deck in front of the sail. Mess dress versions of these badges been identified (GRK-3.3m). These and similar badges were used until around 1951 when Greece became a member of NATO.

GRK-3.3.1

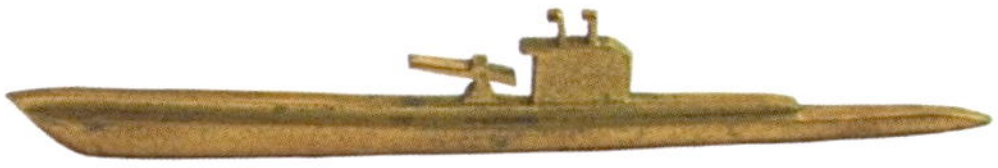

GRK-3.3.2

GRK-3.3m

In 1943, the use of the Royal Hellenic Navy (RHN) Campaign Cross was implemented for sailors serving on ships during WWII. Ribbons were issued with stars for the bulk of the RHN, but submariners serving for more than a year and conducting at least three patrols were issued a black ribbon with a small submarine, gold for officers, silver for NCOs, and copper or bronze for ratings (GRK-CC.1, GRK-CC.2, GRK-CC.3). This was a starboard-facing submarine profile approximately 18mm long, similar to the *Proteus*-class badge profile.

GRK-CC.1

GRK-CC.2

GRK-CC.3

One badge that seems to have created more questions than answers is a small starboard-facing submarine profile with a prominent forward deck gun and a low-profile conning tower, limber holes and a torpedo tube that was used with a stick pin attachment method (GRK-UA1.1?). While some collectors have identified it as a badge worn during WWII, it does not accurately resemble the profile of any of the submarines in use by the Hellenic Navy from 1912 through 1945, and experts within the Hellenic Navy historian community dispute claims that it is a Greek badge. Some sources raise the possibility that it may have been issued as a sweetheart pin during the war. This badge seems to be available in greater quantities than the *Katsonis* or *Proteus* profile badges.

GRK-UA1.1?

Following their entry into NATO in 1951, several submarines were loaned to the RHN by Great Britain, and profile badges from this time feature a more rounded hull (GRK-POSTWAR-UK). These were returned in 1958 and two *Gato*-class submarines were provided to the RHN by the U.S. Navy with corresponding badges featuring a more angular submarine profile (GRK-POSTWAR-US) briefly used until the modern RHN submarine qualification insignia was implemented in the early 1960s.

GRK-POSTWAR-UK

GRK-POSTWAR-US

The badge that ended up becoming the Greek submariner's badge by the 1960s was a starboard-facing submarine profile with a prominent deck gun and a sail with two periscopes, in gilt metal with clutch back attachment (GRK-7). This is earned by officers after five months of submarine training and for ratings after three to five months at submarine school. It is worn on the left chest of all uniforms, and a mini version exists for use by officers and NCOs to wear with their mess dress uniforms (GRK-7m). Multiple variations exist of this badge due to the lack of strict uniform standards.

GRK-7

GRK-7m

A silver badge of the same design was implemented in the late 1970s for junior ratings. This badge was the same size as the gold mess dress badge, but in silver, and worn on the left side of the black neckerchief worn by junior Greek sailors (GRK-8.1 and 8.2). This badge was phased out by 1988.

GRK-8.1

GRK-8.2

In 2010 a Submarine Flotilla Commander badge was approved for use. The same starboard-facing submarine profile of the GRK-7 is used, placed upon two gilt laurel branches forming a "V" (GRK-9). Patches with the submarine badge profile are approved for use on work coveralls for officers with name and rank (GRK-10). To date, no further Greek submarine badges have been approved for use.

GRK-9

GRK-10

INDIA (IND)

The Indian submarine service was born with the commissioning of the INS *Kalvari* on 8 December 1967, the first of four Russian *Foxtrot*-class submarines purchased by the Indian navy over the following two years. The Indian submarine insignia was created shortly after the commissioning of that first submarine.

The Indian submarine qualification insignia is a gilt badge with inward-facing heraldic dolphins on waves, supporting the national emblem of the Republic of India in the center, the Lion Capital of Ashoka (IND-1). Most badges have either a horizontal pin or a lug and pin configuration. While there are multiple slight variations of this badge due to different manufacturers, this design remains the only approved version. Some of these badges have maker's marks on the back, but most do not.

IND-1

This badge is worn by all qualified submariners regardless of rank who successfully complete the submarine board exam. Originally the badge was worn on the right chest, but in the early 1970s it was moved to the left chest due to implementation of a mandatory nametag on the right chest. There is only the gilt (gold) version of this badge. While they exist in the market, there is not now, nor has there ever been any official Indian submarine insignia that is silver. There are also mini gilt badges that exist, but these are used as either tie pins or for use on civilian clothing with no official standing.

There are cloth patches approved for use by Indian submariners on various uniforms, but the authors do not have documented evidence of when the patches were first authorized. They range from the quite simple (IND-1C.1 and 1C.2) to ornate bullion (IND-1C.3). Multiple other variations exist.

IND-1C.1

IND-1C.2

IND-1C.3

In 1986 the first of the Indian *Sindhughosh*-class submarines rolled off the line. At some point some nice quality boat badges were created for at least eight of these submarines. These badges have a laurel leaf and braided rope wreath with the national emblem on top, a port-facing submarine profile in the middle against a blue sky, the name of the submarine in gold letters on a black band sitting under the submarine, and below that the Russian and Indian naval ensigns (IND-BB-1

IND-BB-1

IND-BB-2

IND-BB-3

IND-BB-4

IND-BB-5

IND-BB-6

IND-BB-7

IND-BB-8

through 8). There are a few other boat badges that have surfaced as well, such as this one for the S-42, the INS *Vagli* (IND-BB-S-42), that appears to pre-date the *Sindhughosh*-class badges. Like most boat badges, they have no official standing as a uniform item.

IND-BB-S-42

INDONESIA (INO)

In the mid-1950s as the Indonesian government started to identify conflicts with certain countries as more of a threat, they began laying the groundwork for a submarine force. Negotiating a deal for submarines with the Soviet Union in 1958, two submarine crews were sent to Gdynia, Poland to begin training on submarine skills and tactics. After a year of training, the crews returned home with the first two of an eventual 12 Soviet *Whiskey*-class submarines. The *Korps Hiu Kencana* (the Golden Shark Corps), the official name for the Indonesian submarine force, was launched on 12 September 1959 when these two submarines were officially commissioned into the Royal Indonesian Navy. A conflict with the Dutch in the early 1960s expedited completing the acquisition of the rest of the *Whiskey*-class submarines.

According to submarine badge lore, development of the first Indonesian submarine insignia was accomplished by the submarine crews in Gdynia, Poland for training. Some of them visited an engraver's shop in Łódź and worked with Mr. Alfred Miroslaw to create a badge based on their design. This proposed design was two golden diving sharks supporting a silver port-facing submarine profile. According to official sources, this badge was never formally authorized nor accepted by the navy, however some all brass versions, possibly from this initial effort, have been identified (INO-UA-1).

INO-UA-1

The implementation date of the first approved Indonesian submarine insignia is listed as 21 December 1959. This badge was a gilt laurel wreath with a diving shark in the center, overlaid by a port-facing *Whisky*-class submarine profile (INO-1). It was earned by all personnel regardless of rank and worn on the left chest above the ribbon bar if worn. Due to the lack of expressed design standards, multiple variations of this badge are known.

INO-1

The INO-1 badge was in use until 1975 when the Indonesian navy began transitioning from Soviet *Whiskey*-class submarines to German *209* subs. A new gilt badge was created featuring a more modern slightly starboard-facing submarine bow and conning tower supported on both sides by golden sharks, both atop seven stylized waves (INO-2). The Indonesian *Hiu Kencana* badge is steeped in symbolism:

- Two opposing sharks signify a demonstration of capability and supremacy at sea and a determination to defend Indonesia
- Submarine in combat-ready position represents a potent strategic weapon
- Periscope raised indicates a vigilant attitude in observing every inch of the nation's waters
- Five gills on the sharks signify that submariners breathe in accordance with the Pancasila, Indonesia's five guiding principles
- Seven waves represent the world's oceans

While the INO-2 badge remains the currently approved submarine insignia, Indonesia may hold the record for the number of variations existing of a single badge. The authors have identified over 30 different variants with differences seen in all facets of the badge design – measurements, materials, shades of gold, attachment methods and various aspects of the badge construction. Most versions are hollow back with clutch pin attachment, and without any maker's marks. However, there are solid back versions, and some well-made badges have "*Kapal Selam*" ("Submarine") stamped on the back, and one recent badge with red on the torpedo tubes. Examples shown here for reference.

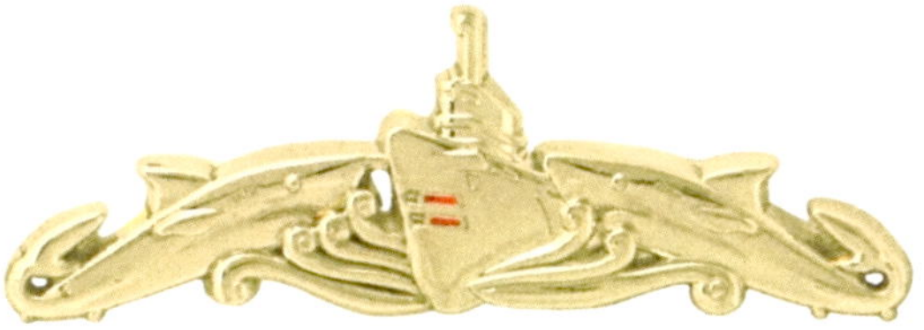

INO-2.1

INO-2.2

INO-2.3

INO-2.4

INO-2.5

INO-2.6

INO-2.7

INO-2.8

INO-2.9

INO-2.10

INO-2.11

INO-2.12

INO-2.13

INO-2.14

INO-2.15

No known patches have been identified for the first Indonesian submarine badge, but like the many versions of the metal badge, the INO-2 cloth patch (INO-2C) has multiple variations. Again, several are shown for reference.

INO-2C.1

INO-2C.2

INO-2C.3

INO-2C.4

INO-2C.5

INO-2C.6

created their own version of the unauthorized German qualification badge. Similar in layout with a port-facing German submarine across the face of an open-topped wreath, it is smaller and much less detailed than the German versions (INO-UA-GER). While the INO-2 badge is worn on the left chest, this badge is worn on the right above the pocket.

INO-UA-GER

Around 2008 the Indonesian naval uniforms changed to incorporate berets for wear with camouflage uniforms. To support this change, submarine beret badges were developed. The first was a gilt metal rectangle with squared off corners on a black leather backing, a naval anchor in the center, surmounted across the middle by the familiar *Hiu Kencana* emblem (INO-3). The early versions of these were relatively crude manufacture with rings attached for sewing them to the beret, and like the badges, manufacturing differences were not unusual (INO-3.1 & 3.2). Later versions were slightly larger with a much more refined and detailed *Hiu Kencana* emblem, three clutch pins for attachment, and the maker's mark for Widjaja Ltd. on the back (INO-3.3).

INO-3.1

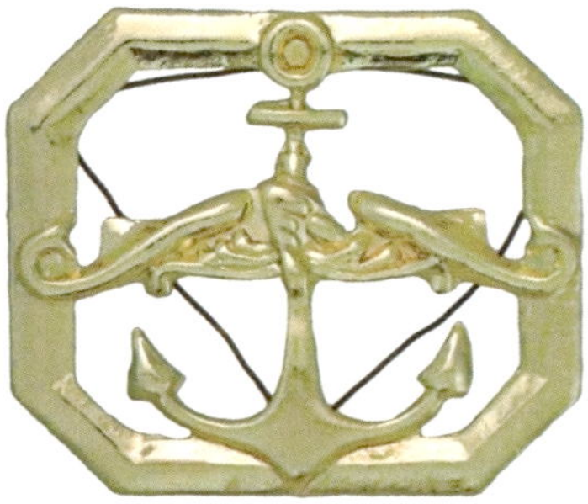

INO-3.2

INO-3.3

In 2023, a new beret badge was issued. For this new badge, many components reversed direction. It is still a gilt badge, but now features a starboard-facing submarine profile, with two sharks rising to meet above the conning tower (INO-4). Apparently, this badge is not very popular with the Indonesian submariners, with many still choosing to wear the previous version, so we will see if it remains in use or not.

INO-4

IMPERIAL IRAN & ISLAMIC REPUBLIC OF IRAN (IRN)

The Imperial Iranian submarine service was initiated in a 1975 agreement with the United States to train Iranian submarine crews on *Guppy*-class submarines they purchased from the U.S. In 1978, Iranian submariners took possession of the Imperial Iranian Ship (IIS) *Kousseh* (*Shark*).

The first Iranian submarine insignia was heavily influenced by the U.S. submarine qualification badge, with inward-facing heraldic dolphins virtually identical to those used on the U.S. badge, a round-hulled bow-facing view of a submarine in the center, with the Pahlevi crown of the Shah of Iran above the sail. They were issued in gold for officers (IRN-1) and silver for enlisted personnel (IRN-2) with a miniature version of the officer badge also made (IRN-1m). These badges were created by Balfour, and original badges have the Balfour maker's mark on the reverse. They were clutch back attachment and worn above the ribbon bar on the left chest of the uniform. There is conflicting information regarding the issuance of these badges. There are reports that they were issued to the crew of the submarine during the ceremony for receipt of the *Kousseh*, and there are first-hand reports from crew members that they never received any insignia at that time. Whichever account is accurate, following the overthrow of the Shah in December 1978, all submarine agreements with the two countries were dissolved, and these badges lost any official standing, if they indeed ever had any. Multiple fakes of these badges exist, including a mini enlisted badge complete with a Balfour maker's mark created after the molds were purchased following the dissolution of Balfour.

IRN-1

IRN-1r

IRN-1m

IRN-1mr

IRN-2

IRN-2r

When the Islamic Republic of Iran began pursuing a submarine force, they turned to Russia, North Korea, and Italy, acquiring multiple midget submarines. In the early 1990s badges began appearing, purporting to be Iranian. These badges were supposedly made in Russia and are a pressed metal badge featuring a port-facing stylized submarine profile, with the national symbol from the state flag captured within a gear wheel that is above the flukes and arms of an anchor, all in relief. The national symbol is stylized representation of the word "Allah" in Arabic script meant to resemble a tulip, Iran's national flower. The examples most commonly seen feature a gold (IRN-3) or a silver (IRN-4) submarine with a red national symbol. It cannot be verified by the authors how the badge was earned, nor where it was worn on the uniform. Like many badges made in Russia, these can be found in large quantities, which of course lends doubts as to their authenticity.

IRN-3

IRN-4

There are multiple known variants of this badge such as those having a blue national symbol on a white background (IRN-3.1). Most of these badges have a screw back post attachment method, but some of the more recent examples have horizontal safety pin attachment methods. The official standing of these badges and the meaning of the variants is not known by the authors at this time.

IRN-3-V1

Since 2020, new badges have appeared with claims of being Iranian. One badge is a gilt port-facing submarine profile flying the country's flag, the Islamic Republic of Iran national symbol pressed in the center in black, and the words "I.R. IRAN NAVY" on the hull (IRN-UK-1). The authenticity of this badge has not been verified by official sources, but there is precedent for other Arabic-speaking countries having words in English on military badges.

IRN-UK-1

Another badge is a gilt port-facing stylized submarine profile with the national symbol pressed into the middle of the sail with a black background that is encircled by a wreath with Arabic wording etched in green at the bottom saying, "Islamic Republic of Iran Navy" and an emblem of three concentric circles in the Iranian flag colors of green, white and red top center. This badge is known in full-size and miniature versions (IRN-5 & IRN-5m). There are known photos of Iranian submariner officers wearing this badge on the right upper chest of their uniform. A similar badge without the wreath is also known with the Iranian Navy emblem pressed into the sail (IRN-5.1). The qualification process to earn these badges is not known by the authors.

IRN-5

IRN-5m

IRN-5.1

There is also a new gilt badge that is the same style as IRN-3 and IRN-4, but with a red national symbol in the middle on a bright blue background and the green, white and red of the Iranian flag horizontally across the sail (IRN-6). A smaller, possibly mess dress version with the same profile and the Iranian flag colors on the sail exists, but the national symbol in the middle of the gear is shallowly engraved on the gilt hull and is not in relief like the previous versions (IRN-7).

IRN-6

IRN-7

According to sources, officers and men supposedly wear the same badges, so the number of presented badges seems to be too large for all of these to be valid. With the existing uncertainty regarding the requirements for earning them and what each badge means, it is possible that several of these badges may be fake, however, they are presented here until compelling evidence is provided verifying validity.

ISRAEL (ISR)

About 10 years following the formation of the state of Israel, the military received approval to create a submarine force. In 1956 the process was initiated to purchase two submarines from Great Britain, which included training the crews. In December of 1959, brand new Israeli submarine qualification badges were presented to the officers and crew of the INS *Tanin* when it pulled into Haifa, Israel. Six months later, the same badges were presented at sea to the officers and crew of the *Tanin's* sister ship, the INS *Rahav*. This first badge design was a silver port view of a submarine with a rather flat profile, showing double saddle tanks, limber holes on the bow, and a snorkel and periscope. This was overlayed with the symbolic Israeli military short sword and an olive branch above anchor arms (ISR-1). This badge was worn on the left breast of all uniforms.

ISR-1

Unfortunately, these badges were not well received by the officers and men of the two submarines. So, by 1961, a new badge had been designed and was issued to all qualified submariners, with the same badge again being issued to both officers and crew. This was a less detailed and more stylized and rounded version of the original badge, but having the same anchor, short sword and olive branch (ISR-2.1 & 2.2). This badge remains the approved initial submarine qualification badge for Israeli submariners to the current day and is issued to personnel completing the rigorous submarine qualification process. This badge can be permanently worn on all military uniforms following three years of submarine service.

ISR-2.1

A "Senior Submariner" badge was approved for use sometime around 1970 to designate 10 years of submarine service. This is the same silver qualification badge, but with a wreath and a centered six-pointed star added and arched over the sail (ISR-3.1 & 3.2). The attachment method for the ISR-1 and also older versions of the ISR-2 and ISR-3 was either a single or double screw-back posts. Newer badges use clutch back attachment.

ISR-3.1

There can be some marked differences in badges depending on manufacturer. The arms in the anchor for the ISR-2 & ISR-3 badges can be either pierced or solid, as can the area between the wreath and the sail for the senior submariner badge. Also, the saddle tanks on the ISR-2 badges can be either flat or more rounded with the ends blending into the hull. There are no known hallmarks associated with either the ISR-1, 2 or 3 submarine badges, however, the ISR-2 and ISR-3 badges that are initially presented to Israeli submariners are serialized, are a slightly heavier construction, and have pierced anchor arms. Subsequent spare badges obtained by qualified personnel are not serialized.

ISR-2.2

ISR-3.2

These badges can be worn with a blue background for peacetime service, but sources report that many submariners choose to not wear the blue backgrounds. A red background is authorized for wear if the submariner conducted combat patrols, which, to date, was only ever authorized once during the Six Day War in 1967.

A miniature badge was approved at a later date for personnel undergoing submarine training (ISR-4). This is a smaller version of the ISR-2 submariner initial qualification badge and is worn as a designation badge until the trainee passes submarine training and is qualified to wear the full-sized insignia. A miniature Senior Submariner badge exists, as does an even smaller version of the basic qualification badge (only 28mm long). Neither of these badges are approved for wear in an official capacity, and can be found worn on civilian clothing or as tie tacks.

ISR-4

In the mid-1990s, a special badge was issued for Unit 663, the electronic and computer intelligence group in the 7th Submarine Division specializing in underwater reconnaissance. This badge is a much smaller version of the starboard-facing submarine profile found on the ISR-2, with the arms and flukes of a naval anchor framing the bottom and a fleur-de-lis atop the conning tower (ISR-5). Genuine versions of this badge have the maker's mark for the company Zehovoy in Hebrew on the back.

ISR-5

ISR-5r

There are individual boat badges, originally issued in the early 1990s for the seven submarines in the Israeli fleet. Very colorful and eye-catching, these are not considered to be qualification badges, so do not have any official capacity and are often worn with civilian clothes (ISR-BB-1 through 7).

ISR-BB-1 *Leviathan*

ISR-BB-2 *Dakar*

ISR-BB-3 *Gal*

ISR-BB-4 *Tanin*

ISR-BB-5 *Thekuma*

ISR-BB-6 *Rahav*

ISR-BB-7 *Dolphin*

ISR-BB-1

ISR-BB-2

ISR-BB-3

ISR-BB-4

ISR-BB-5

ISR-BB-6

ISR-BB-7

Two other badges are worth talking about that never achieved issuance. Shortly after the original 1959 badge was issued, a badge with the same design but with a single star atop the sail was approved to designate senior submariners. Later, a gold badge, similar in design to the senior submariner badge with a wreath and a star was approved for Commanding Officers. Though both were approved, neither of these badges were ever put into production. However, unauthorized versions of both exist.

ITALY (ITL)

With a history of and national pride in shipbuilding since ancient times, it is not surprising that Italy was one of the early adopters of submarines, with the *Delfino* being their first submarine commissioned into the *Regia Marina* (Royal Navy) in 1896, though the Italian submarine service dates their anniversary from the *Delfino's* launching in 1890. The Italians had a fleet of submarines before most countries had their first boat, having 21 submarines by the start of WWI.

While sources identify that it is possible a submarine badge was created for submarine enlisted personnel as early as 1916, the authors have been unable to uncover any evidence supporting this. What has been verified is that on 16 January 1918, Act #36 was issued in the *Regia Marina* Official Bulletin authorizing a Trades insignia for submarine crews. This was a round silver badge 46mm in diameter with a silver leaping dolphin facing right in the middle and the word "SOMMERGIBILI" on the top half (ITL-1). It was issued only to Petty Officers and ratings personnel. Worn on the left sleeve above the specialty and rank insignia, very early versions were fastened by being sewn on through two holes 180° apart or by use of a vertical pin. Later versions have a small horizontal pin attachment behind the dolphin. This badge is still the Italian enlisted submariner badge, and as such is the oldest existing submarine insignia design still in use, with only a brief two-year hiatus in use from 1939-1941. There are period examples of this badge that appear to be gold, however this is due to the base material absorbing the silver washing as the badges aged. No gold versions of this badge were made.

ITL-1

A few years later, the Royal Navy decided to include the officers when Act #491 was issued on 25 September 1924, authorizing a new badge for officers and petty officers while ratings personnel retained the original badge. The new badge also utilized a dolphin, but was much smaller. This badge is a 15mm diameter ring with the leaping dolphin in the middle, this time facing the wearer's left, and "SOMMERGIBILI" arcing over it, and topped by the Savoy Crown. It was gilt (gold) metal for officers (ITL-2.1, 2.2, & 2.3), silver metal for petty officers (ITL-3.1 & 3.2), and was to be worn on the left breast, 10mm above the ribbon bar, or in place of one. There were multiple variants of these badges made by different manufacturers with slightly different dimensions, crowns, and attachment methods, as well as a slightly smaller mess dress version of the ITL-2 with a 12mm diameter ring (ITL-2m*) that often appears with a button-hole attachment point.

ITL-2.1

ITL-2.2

ITL-2.3

ITL-3.1

ITL-3.2

Wire bullion versions of each (ITL-2C & ITL-3C) were manufactured in multiple variations that were sewn in the proper location on the uniforms.

From the 1910s through 1930s, the Italians were one of the leading exporters of submarines in the world, providing them to multiple countries, including to the United States. It is important to note that during this time Italian badge manufacturers also provided badges (typically boat profiles) to other countries including Brazil and Spain.

ITL-2C

ITL-3C

In 1939 the use of the petty officer (ITL-3) and rating badges (ITL-1) was suspended, and all personnel were allowed to use the gold officer's badge (ITL-2). Apparently, this decision was not well-received, because on 31 March 1941, Act #131 once again authorized the use of the ITL-1 original sleeve badge for ratings personnel, while making the ITL-2 badge the only badge used by both officers and petty officers.

An order dated 11 November 1941 authorized awarding a special "Honor Badge" with an accompanying certificate to those submariners serving aboard for over four years. The time requirement was lowered to three months for serving in combat during the war. It was also issued to personnel who had been removed from serving aboard submarines due to wounds, and to civilians acting in their capacity aboard the subs that met the same requirements as the sailors. This badge was a similar design to the ITL-2 in that it was gilt metal, round with "SOMMERGIBILI" arcing over the leaping dolphin, and topped with the Savoy Crown, however, it is larger with a 25mm diameter, and instead of a simple ring, the circle is a laurel wreath (ITL-4.1). The majority of these are based on a design by Picchiani & Barlacchi, and many examples of the officially presented Honor badges during WWII have the word "DEPOSITATO" on the back. Other variations are known with red enamel in the crown (ITL-4.2), and a rare hollow-back version designed by Stefano Johnson with red enamel in the crown (ITL-4.3).

ITL-4.1

ITL-4.2

ITL-4.3

After Mussolini was replaced as Italian Prime Minister in July of 1943 and Italy signed an armistice with the Allies, within a few months the *Repubblica Sociale Italiana* (RSI), or the Italian Social Republic, was established in northern Italy. While the official Italian navy continued to use the previous badges, members of the RSI removed the Savoy crown from their qualification and Honor badges because they no longer acknowledged the monarchy. Some badges were manufactured without the crown as well, effectively making them similar to badges used post-war.

Another type of qualification badge created during the war was the Italian Navy "Long War Navigation Badge." These badges were created for all naval units, with the only difference being the figure in the middle. The submarine version was a vaguely cross-shaped badge framed by either a rope or a solid tubular boundary with a Savoy crown atop, a naval anchor on the inside supporting the frame, and a port-facing torpedo mounted by shark in the middle. There were three classes - 1st class bronze, requiring at least 18 months aboard the sub, or 1,000 hours of navigation with at least one combat action (ITL-5.1), 2nd class silver, requiring at least 30 months aboard, or 3,000 hours of navigation with at least three combat actions (ITL-5.2), and 3rd class gold, requiring 48 months aboard, or 5,000 hours of navigation with at least 6 combat actions (ITL-5.3).

ITL-5.1

ITL-5.2

ITL-5.3

Submariners had requirements put in place to expedite the manner by which they could earn the necessary hours to earn these badges, with each hour at sea counting as 1.5 hours toward the requirement, and any combat action reported on the official war bulletin counted as two hours. As with the other badges, the RSI also issued these in crownless versions that were also used post-war and are still manufactured (ITL-5.1NC, ITL-5.2NC, and ITL-5.3NC).

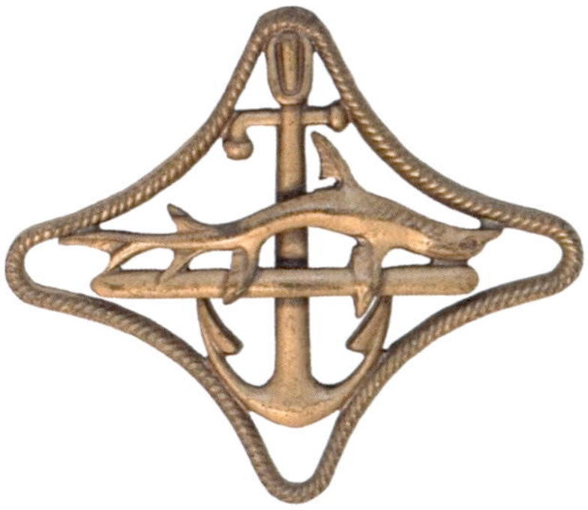
ITL-5.1NC

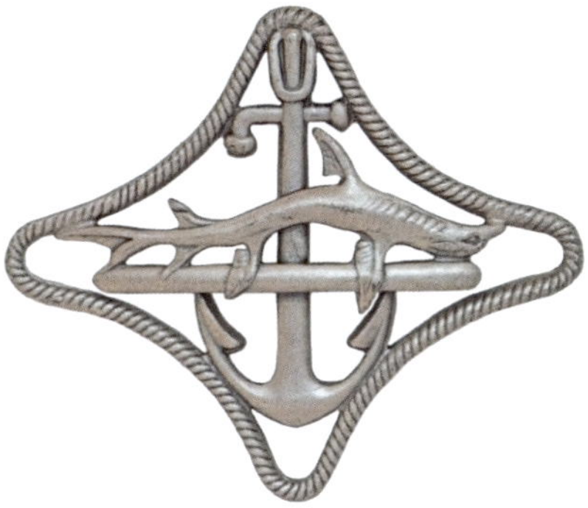
ITL-5.2NC

ITL-5.3NC

During the war many variants of Italian qualification badges were produced. One of these was created by Italian submariners stationed at a base in Bordeaux, France called "BETASOM" from which Italy conducted numerous submarine operations in the Atlantic Ocean. The personnel in the submarine squadron there attached a red enameled letter "A" over the ITL-2 badge to designate "*Atlantico*" for the Atlantic Ocean operations (ITL-2-UA-BETASOM.1 & .2). This was not recognized by the brass as official but was tolerated for wear.

ITL-2-UA-BETASOM.1

ITL-2-UA-BETASOM.2

Near the end of the war the personnel serving in the Atlantic submarine bases in France created the *Divisione Atlantica* sleeve badge. This was a rather large metal shield with the Italian submariner dolphin curled about a Roman short sword and facing to the wearer's left. Examples are known in silver similar to the copy shown here (ITL-DA-SLV.1), and also with blue paint or enamel (ITL-DA-SLV.2). Valid examples of this badge are rare, and many copies exist.

ITL-DA-SLV.1

ITL-DA-SLV.2

There were several other badges created for use by various organizations such as *Decima Flottiglia MAS* and the battalion of ex-*Scire* submariners as well as literally dozens of finely crafted boat badges (ITL-BB), mostly produced by Picchiani and Barlacchi or Stefano Johnson. There was also submarine qualification badge jewelry such as cufflinks (ITL-P&B-CUFF) and tie tacks produced in significant quantities. These are all beautiful pieces, but we will not cover them here as there are several comprehensive Italian books and magazine articles on this subject as noted in the reference section.

ITL-BB

ITL-P&B-CUFF

Following WWII, a vote was held on 2 June 1946 that resulted in the dissolution of the Italian monarchy and officially creating the Italian Republic. Subsequently, all symbols of the Savoy monarchy were removed from official insignia. Initially the Italians were forbidden operational submarines, however, they managed to hang on to two that were used only as mobile electric generators. This changed in 1952 when those same two were put into service as training submarines. After Italy joined NATO in 1954, the U.S. Navy gave them five submarines, but the Italian submarine-building machine took off again in the mid-1960s when they started building submarines again after a 20-year hiatus.

During this post-war period, the ITL-1 remained the qualification badge for enlisted submariners and the de-crowned officer and petty officer's badge became the service badge for the officers and chief petty officers (ITL-2NC). The de-crowned Honor badge (ITL-4NC) now became a badge issued only to officers or chief petty officers with at least five years of service aboard submarines. These ITL-4NC badges were available in several styles, with examples based on the Picchiani & Barlacchi die (ITL-4NC.1) and on the Stefano Johnson die (ITL-4NC.2) even though both were produced by multiple manufacturers. Silver versions exist that are not official badges.

ITL-2-NC

ITL-4NC.1

ITL-4NC.2

These officer and chief petty officer badges remained in place until 1991 when the rules changed allowing crowned badges again and new badges were issued. The qualification and Honor badges were both modified to include the Emperor Claudius "Naval" crown (ITL-6 and ITL-7.1 & 7.2), and a mess dress version of the service badge was also approved (ITL-6m). Versions made by different manufacturers exist for these badges.

ITL-6

ITL-6m

ITL-7.1

ITL-7.2

When the rule change was issued in 1991 allowing the crowns, a manufacturer misinterpreted the criteria for the new badges and issued a service badge (ITL-6-WC*), an Honor badge (ITL-7-WC), and a mess dress badge (ITL-6m-WC*) all with the Wall crown, and not the Naval crown. The Wall crown is typically reserved for use on Italian Army or Air Force insignia. Though this was initially considered an error, sources identify that the Honor badge with the Wall crown has been used for the blue "Basque" caps worn by submariners since that crown is typically used for cap wear.

ITL-7-WC

Italian submariners were authorized to wear berets in 2020, and with this a newly authorized beret badge was issued. With a design similar to the Honor and service badges with the leaping port-facing dolphin and outer wreath, it is significantly larger than the qualification badges with a 49mm diameter. It is issued in bronze for all crew members (ITL-8.2), and in gold for commanding officers (ITL-8.3). The first beret badges issued in bronze were cast (ITL-8.1), so were quite fragile, resulting in crew members requesting a more durable version. New versions were made that were die-struck and more robust. The original beret badges were attached with two clutch pins, and the newer version is attached with a lug and pin configuration. This is the last variant of official Italian submarine insignias issued to date, but with their long history with both submarines and art, it is likely we will see more in the future.

ITL-8.2

ITL-8.3

ITL-8.1

IMPERIAL JAPAN & JAPANESE MARITIME SELF-DEFENSE FORCE (JPN)

The Imperial Japanese Navy's first submarines were several *Holland*-class submarines purchased from the Electric Boat Company in the United States in 1904. These were assembled and ready for operation in Japan by 1905 for the Russo-Japanese War, but the submarines never conducted any war patrols. This purchase provided the impetus for the initial effort to implement a submarine qualification badge a few years later.

Notice No. 4 issued 17 April 1909 by the Japanese Navy gave the description of the newly proposed Navy Submarine Service badge. Approved by the Emperor, this badge was a maroon painted bronze *Sakura*, or cherry blossom, with a starboard-facing *Holland*-class submarine profile made of a "silver-plated" metal in the center (JPN-1.1). Issued to all petty officers and enlisted personnel after two years of assignment to a submarine, the badge was worn on the lower right chest of dress uniforms, with a vertical clasp on the rear providing attachment. The notice specifically stipulated the award was to be "returned to the government" by the recipient when no longer attached to a submarine.

Two years later, Notice No. 9 was issued on 8 September 1911, making minor specification modifications to the badge, changing the base metal of the lotus flower to copper, and making the submarine profile "silver colored" (JPN-1.2).

JPN-1.1

JPN-1.2

In 1915, an effort was mounted to create a submarine Commanding Officer's badge, however, this failed due to opposition within the submarine force. The proposed design was a vertical oval rising sun in alternating matte and shiny silver, a gold submarine bow-on profile in the center with bow waves curving up the sides of the badge (JPN-PROP-CO). The opposition to the CO badge was due to the belief that engineering officers also required special training, but without similar acknowledgement.

JPN-PROP-CO

While the submarines were originally based in Yokosuka, it was determined that the seas in the area were too rough to allow for adequate training, so they were moved to Kure and the submarine school was established there on 3 September 1920. It was at this point the badge became known as the Navy Submarine School Trainee Graduation badge, issued to petty officers and enlisted personnel who completed the course. Only personnel with duties that were "essential to submarine warfare" attended the school, so not all members of the crew received the badge. Enlisted personnel who were promoted to warrant officer had to surrender the badge. This badge was in use by the Imperial Japanese Submarine Force with certain variations until the 31 July 1944 when it was abolished, retroactive to 1 April of that year.

Following WWII, Japan faced several challenges for reestablishing a submarine force, most notably the resistance of a now very anti-military Japanese public to submarines as aggressive weapons of war. So, when Japan finally purchased a submarine from the United States in 1955 for training their anti-submarine warfare ships, they chose to classify it as an "underwater high-speed target". As a means to ensure adequate country defense, the Japanese Maritime Self-Defense Force (JMSDF – established in 1954) finally approved construction of new submarines, with their new *Oyashio*-class submarine being launched in May 1959 and commissioned in June 1960.

The JMSDF quickly started construction of several other submarines and began ramping up their submarine force personnel as well. By 1962, with several new submarines rolling off the line, "Maritime Self-Defense Force Instruction No. 27 – Instruction for Aviation Badge and Submarine Badge" was issued outlining the requirements for a new Japanese Submarine Force qualification insignia.

This new badge had inward-facing heraldic dolphins separated by a *Sakura* surmounting a vertical naval anchor with stylized waves connecting them. This badge was issued in gold for officers (JPN-2.1 & JPN-2.2) and silver for enlisted personnel (JPN-3). The regulations specify the badge is to be 60mm long and 20mm high. It is surmised that early badges were serialized due to multiple known examples existing, however, documentation verifying this is not known to the authors.

JPN-2.1

JPN-3

JPN-2.2

JPN-2.2r

At some point, badges of the same design, but slightly longer (69mm) began to be manufactured. This was not due to a change in design regulations and is thought to be only due to a different manufacturer (JPN-2L and JPN-3L). There are multiple variants of both styles of these badges, with minor differences in dolphins, *Sakura*, anchor, waves and attachments points with clutch back, horizontal pin, and threaded posts known. While it is assumed these differences are due to different manufacturers, there are no known examples with any maker's marks on the back.

JPN-2L

JPN-3L

In 2006 a badge was issued for Japanese submarine medical officers. This badge is similar in size and design to the JPN-2 badge, however the anchor has been replaced with a shield bearing a caduceus with the *Sakura* mounted above (JPN-4.1 & JPN-4.2). This badge is issued to Navy doctors who complete an additional course on submarine medicine and who spend at least two years in work related to the health of divers and submarine crews. No known examples of this badge exist in the long version.

JPN-4.1

JPN-4.2

Patches for officers and enlisted personnel are used on submarine coveralls and designs are known with the profiles of both the shorter and longer metal badges and have been in use since the mid-1990s. These exist in thread and bullion variants with black, blue, and green backgrounds, some with JMSDF printed on them (JPN-2C & 2LC & JPN-3C & 3LC).

JPN-2LC

JPN-2C

JPN-3LC

JPN-3C

LIBYA (LIB)

Between 1976 and 1983, the Libyan government obtained six *Foxtrot*-class submarines from the USSR. Various sources report that the submarine force conducted no patrols past the mid-1980s with reports of submarines sinking due to neglect or lost due to sanctions over the years. The last reported Libyan submarine was seized by rebel forces at the Benghazi naval base during the Libyan civil war in 2011.

Though details for development of a submarine qualification badge for Libyan submariners are slim, it is widely believed that the crews received the qualification badges around 1976, and very possible that the badges were made in the USSR. It is generally accepted that the approved version is a starboard-facing profile of a Foxtrot submarine in gold (LIB-1) with photos of senior Libyan officials wearing the badge seemingly confirming this. Reproductions of these badges far surpass the number of known valid examples. The authors do not have information regarding if these badges were issued to only officers (following the Soviet protocol) or all personnel on board, the process for earning them, nor in what manner they were worn.

LIB-1

A cloth badge featuring a gold bullion rounded starboard-facing submarine profile (LIB-2?) with a horizontal pin back has been thought to be a Libyan submarine distinctive by researchers since the 1980s, however the authors have found no valid information supporting this claim.

LIB-2?

MALAYSIA (MAL)

As early as the 1980s the Malaysian military was sending officers to multiple countries for familiarization with submarine operations, but it was not until the mid-2000s that serious efforts were made to obtain submarines. Their efforts finally came to fruition with the purchase of two *Scorpène*-class French submarines that were commissioned in 2009, with the country's new submarine insignia being proudly displayed by the commissioning crews.

This gold metal badge has a centered fouled naval anchor topped with the Malaysian royal crown with the words "Allah" on the left and "Mohammed" on the right in the red enamel of the crown. The anchor is supported by inward-facing leaping bottle-nosed dolphins, and the space between the anchor arms and the dolphins is vacated. This badge (MAL-1.1) is clutch back with the maker's mark "ELM" for Eng Leong Medallic Industries. A different version of this badge was also made by ELM with the same design as the first badge, but without the maker's mark (MAL-1.2).

MAL-1.1

MAL-1.1r

MAL-1.2

MAL-1m

This is a qualification badge that is earned through a five-level qualification course that culminates with the submariner candidates successfully completing a board chaired by the Deputy Chief of the Navy. It is worn by all submariners, regardless of rank, and after being earned can be worn even after leaving the submarine service. It is worn on the upper right breast of all uniforms. A mess dress version is approved (MAL-1m), and patches are authorized for work/underway uniforms – one in blue, and the other in olive green (MAL-1C.1 & MAL-1C.2).

MAL-1C.1

MAL-1C.2

Another newer version of the metal badge is much thicker than the previous versions, with more texture and detail on the anchor, colored "jewels" on the brim of the crown, and the space between the dolphins and the anchor arms is solid. There is no maker's mark on this badge, and it is also clutch back (MAL-1.3).

MAL-1.3

MAL-1m.3

MYANMAR (MYA)

Myanmar began the process of establishing a submarine fleet as early as the mid-2000s, however, no progress was made toward this goal until October of 2020 when they completed the acquisition of a *Kilo*-class submarine from the Indian navy. A little over a year later, a Chinese submarine was added to the Myanmar fleet. At some point during this process, a submarine badge was created, but no details on specific dates have yet been revealed to the authors.

The original badge is a gold naval anchor in front of a globe with a five-point star on top, flanked by two inward-facing bottle-nosed dolphins. The thin gilt badge appears to have been hammered, and is backed by an accompanying cut-to-fit piece of black acrylic (MYA-1.1). According to Myanmar authorities, this badge is worn by officers only, with no insignia for enlisted personnel.

MYA-1.1

In early 2023, a more refined version of this badge appeared. It is the same design but is a substantially thicker die-struck badge with the same matching black acrylic backing, but with smaller dimensions (MYA-1.2). It is not known if this badge is approved by official sources, or is an unsanctioned copy of the original, as it appears to be available in larger numbers.

MYA-1.2

There is also an authorized patch insignia, with the badge design represented in gold thread on a black background (MYA-1C).

Not much is known about the process by which this badge is earned/presented, however, sources claim that this is a qualification badge, so the process may be similar to that used in the Indian Navy.

MYA-1C

NETHERLANDS (NET)

As one of the earliest adopters of submarine technology, the Royal Netherlands Navy Submarine Service, the *Onderzeedienst* (OZD), was established with the commissioning of their first submarine on 21 December 1906. HNLMS *O-1* was soon joined by others, and the country had more than 20 submarines by the end of WWI. Despite being on the leading edge of submarines, the Dutch were lagging in adopting a submarine insignia other than a cap tally.

According to sources, the first attempt at a submarine-specific badge by the Dutch Navy was when a *torpedist-onderzeedienst,* or "submarine torpedoist" trades badge was created for submarine torpedomen in 1935. It was approved for wear on both the dress blue (NET-TORP.1) and tropical white uniforms (NET-TORP.2) as these replicas show. It is not known how long this badge was in use.

NET-TORP.1

NET-TORP.2

A badge exists in the Dutch Marine Museum that may be the Netherlands' first attempt at an all-hands submarine insignia. According to the museum the badge dates from near the end of WWII and appears to show a bow-facing submarine angled to starboard riding in waves and surmounting two crossed torpedoes (NET-PTYP?). There is no documentation proving or disproving this as an initial submarine insignia effort.

NET-PTYP?

It wasn't until the commissioning of four *Dolfijn*-class submarines in the early 1960s that we have the first documented effort by the Dutch Naval Command for an official submarine insignia. Two similar designs were accepted for consideration, and prototypes of each were manufactured and distributed for consideration. The basic design for both was a gold badge with twin bottle-nosed dolphins facing away from each other, their tails encircling a bow-facing submarine. What made them different was the first prototype (NET-P1) had more stylized dolphins, with thinner tails, bulbous bodies, and short, stubby bills. Approximately 50 of these badges were distributed for the trial.

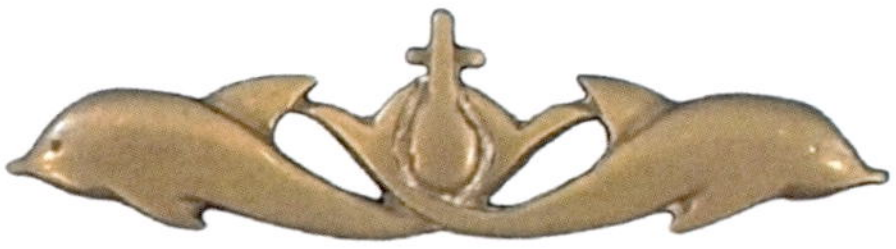

NET-P1

The version that was accepted had dolphins that were more proportional and were considered more graceful and beautiful by the submarine force personnel (NET-1). It was overwhelmingly selected as the winner and was certified as the OZD qualification insignia by official Naval decree #694937on 24 February 1965. Shortly after, the very first Dutch qualification insignia was presented on 23 April 1965 to the Commander Dutch Submarine Forces, Captain Woldering.

NET-1

Though there are some minor differences between manufacturers such as thickness, the badge has remained unchanged since 1965. Some early versions had a horizontal pin attachment method (NET-1.1), but the clutch pin attachment is by far the most common. The badge is worn below the ribbon bar on the left chest.

NET-1.1

NET-1.1r

The Dutch dolphins, called “Flipper” by the OZD personnel, are earned through a rigorous approximately year-long qualification process. It is a two-step process where a theoretical learning section is completed prior to the trainee beginning the practical part of the qualification process that includes systems, watch standing, and drills. Upon successful completion of a final written exam, the sailor is considered to be qualified on submarines, and can proudly wear Flipper.

There is a mini version of the Dutch badge that is typically seen in the form of a stickpin, however, this is unofficial and is not worn in any official capacity, as the full-sized dolphins are also worn on the mess dress uniform.

Around the late 1980s, blue cloth qualification insignia patches were approved for use on work uniforms (NET-1C). Similar patches in olive drab and desert tan were proposed for use, but neither were approved for official use, nor ever used, and only a small number exist.

NET-1C

Though not qualification badges, in 1962 the Dutch Submarine Command issued two commemorative badges to the crews of the Dutch submarines *Dolfijn* (NET-COM1) and *Zeeleeuw* (NET-COM2) to celebrate their completion of an around the world cruise. There were only 50-60 of each badge made for the members of the crews.

NET-COM1

NET-COM2

With the recent decision to replace their aging fleet with four new diesel-electric submarines, the Dutch Submarine Force shows no sign of slowing down, so the possibility exists for additional badges in the future.

NORWAY (NOR)

Despite having a continuous submarine force presence since 1909, it was not until 1958 that the first Norwegian submarine qualification insignia other than a cap tally was introduced. Initially, a gold badge marked, "830S OTTAR HVAL" on the back top center for maker and silver content was worn by all qualified personnel. This badge is a starboard-facing stylized submarine with two scaled mythological fish flanking the sail and the Norwegian crown in the middle above them (NOR-1-OH).

NOR-1-OH

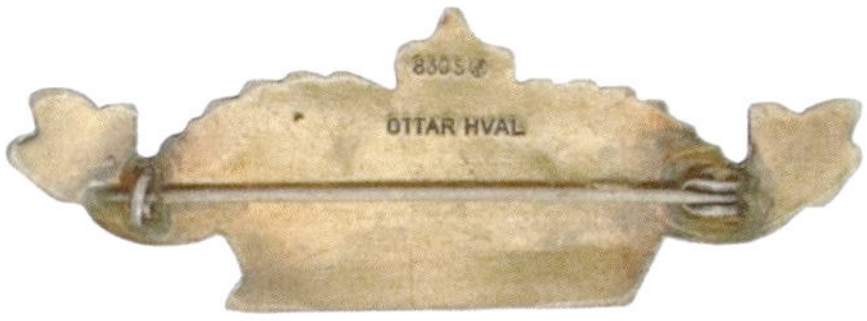

NOR-1-OHr

In the early 1960s, a bronze "Junior Submariner" badge was introduced (NOR-2-OH), and the gold badge became the "Senior Submariner" badge. These badges remained essentially unchanged for four decades. While slight variations exist, these older badges all had horizontal pin attachment points.

NOR-2-OH

NOR-2-OHr

The Junior Submariner "Submarine Wings" are earned by personnel upon completion of submarine training and after they pass the bronze badge qualification test and qualify as a watch stander during their first 6 months aboard a submarine. The Senior Submariner badge is earned by those going on to serve as part of a submarine crew for at least 12 months after passing the more advanced gold badge qualification test. This process is the same for both officers and enlisted personnel. The badges were originally worn on the left chest of all uniforms, above the ribbon bar when present, but were switched to the right side in the mid-1960s.

In the early 1980s, the manufacturer David Anderson took over production from Ottar Hval (NOR-DA.1). All Ottar Hval and David Anderson badges were produced used the same original die from 1958 and can be identified by a small linear die flaw on the lower left side of the crown.

NOR-1-DA.1

NOR-1-DA.1r

There were a few other manufacturers starting in the 1980s, notably, "LCK" (NOR-1-LCK & NOR-2-LCK) and others that were unmarked.

NOR-1-LCK

NOR-1-LCKr

NOR-2-LCK

NOR-2-LCKr

A full-sized gold bullion badge was introduced in 1959 (NOR-1C.1). These were worn through the 1980s but are now obsolete. Gold bullion miniature badges (NOR-1Cm) were worn for mess dress uniforms from 1959 until being replaced with a metal miniature badge made by Sporrong with the new style crown in the 2010s (NOR-1m-SPR). There was a brief attempt in 1996 to revive the full-size gold bullion badges, but the issued version was only in use for a short period prior to being discontinued (NOR-1C.2). One proposed version with a red crown was rejected, but several copies were distributed (NOR-1C-REJ).

NOR-1C.1

NOR-1Cm

NOR-1C.2

NOR-1C-REJ

In the mid-1990s the crew of the *Skolpen* ordered some tie clips in a special design from a local badge manufacturer. The design was the same as the qualification badge in gold but was framed on a black background. The manufacturer misunderstood the order and delivered 100 full size badge tie clips instead of the tie clips with a miniature badge as requested (NOR-SKOLPEN-FS-TC). When the crew complained, the manufacturer resolved the issue and delivered the correct sizes (NOR-SKOLPEN-MIN & NOR-SKOLPEN-MIN-TC) along with 100 full size badges of the same design (NOR-SKOLPEN-FS). The badges are metal with clutch pin attachment and are not authorized for official use.

NOR-SKOLPEN-FS-TC/FS

NOR-SKOLPEN-MIN

NOR-SKOLPEN-MIN-TC

Special commemorative badges were issued by David Anderson in 1999 & 2009 to celebrate the 90th and 100th anniversaries of the Norwegian submarine service. These badges were also created using the original die from 1958, with 100 badges created for the 90th anniversary (NOR-1-DA-1999), and 200 being created for the 100th (NOR-1-DA-2009).

NOR-1-DA-1999

NOR-1-DA-1999r

NOR-1-DA-2009r

In the late 2000s, some new style badges began to appear with slightly more stylized crowns and sea creatures, some even with clutch back pin attachments, including one by manufacturer Per Hauge who put out a limited run of badges with a textured back due to a distribution issue with the regular provider (NOR-1-PER).

NOR-1-PER

NOR-1-PERr

Others during this time included Swedish badge manufacturer Sporrong now making badges for the Norwegian military (NOR-1-SPR.1, NOR-1-SPR.2, NOR-2-SPR.1).

NOR-1-SPR.1

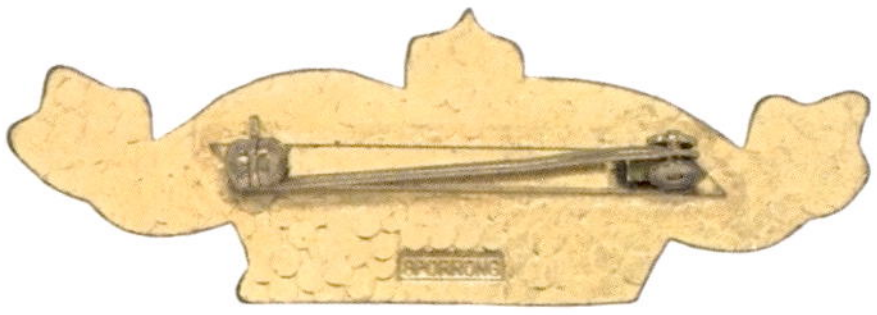

NOR-1-SPR.1r

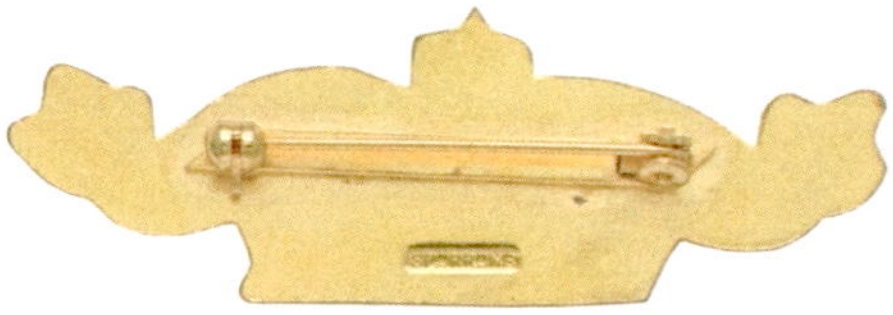

NOR-1-SPR.2r

NOR-2-SPR.1

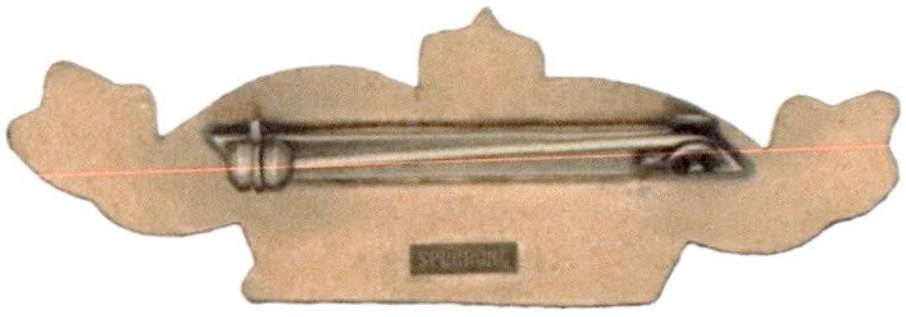

NOR-2-SPR-1r

However, it was not until the late 2010s that a revised badge with a new style crown was issued. These were typically hollow back, had a wider crown, and were predominantly made by Sporrong. These were issued in gold (NOR-1-SPR.3) and bronze (NOR-2-SPR.2) with miniature badges also being issued for the senior submariners (NOR-1m-SPR).

NOR-1-SPR.3

NOR-1-SPR.3r

NOR-1m-SPR

NOR-1mr-SPR

NOR-2-SPR.2

NOR-2-SPR.2r

In the late 2000s, a Submarine Service Branch badge was introduced in full size (NOR-3) and miniature (NOR-3m). This is not to indicate qualification, but is a designation badge used to show the branch of naval service. As such, it is required to be worn on the right side of the uniform, which had the effect of moving the submarine qualification badge back to the left side where it was originally.

NOR-3

NOR-3m

In 2021 David Anderson released another badge created from the original Ottar Hval die, maker marked with "D-A925S", which was the same maker's mark that was used for the 100th anniversary commemorative badges. This badge (NOR-1-DA.2) was ordered by the submarine officer's club "The Periscope" for sale to qualified submariners who want a higher-grade gilded sterling silver badge in the original design from 1958, rather than the current regular issue badge.

NOR-1-DA.2

NOR-1-DA.2r

In the 2010's cloth patches for the submarine qualification badge and the submarine service branch badge were approved for the sea and camo rigs. These were created in white on dark blue (NOR-1C.3 and NOR-3C.1), black on olive drab (NOR-1C.4 and NOR-3C.2), and brown on desert tan (NOR-1C.5 and NOR-3C.3).

NOR-1C.3

NOR-1C.4

NOR-1C.5

NOR-3C.1

NOR-3C.2

NOR-3C.3

The latest version of the Norwegian qualification badges appeared around 2021 and are very shiny badges that are more similar to the David Anderson/Ottar Hval version than the Sporrong version with solid backs and clutch pin attachment (NOR-1-UK & NOR-2-UK),. There is no known manufacturer for these badges. These and all previous versions are still worn since they are approved for use.

NOR-1-UK

NOR-1-UKr

NOR-2-UK

NOR-2-UKr

PAKISTAN (PAK)

Pakistan began their submarine service with the leasing of a *Tench*-class submarine from the United States in 1963 that was commissioned into the Pakistani Navy in 1964. Shortly after receiving the submarine, the Pakistani Navy wasted no time in implementing a submarine insignia.

The Pakistani submarine qualification insignia was issued in gold for officers (PAK-1) and silver for enlisted men (PAK-2). The badge is a very simple profile of a stylized starboard-facing submarine with a wreath encircling the center of the hull and the sail that is topped with the Islamic star and crescent moon. Multiple variations of this badge exist due to various manufactures and lack of strong standardization requirements, however the overall design has stayed essentially the same. There is a miniature version of the officer's badge (PAK-1m) that is worn on mess dress uniforms for official events and formal dinners.

PAK-1

PAK-2

PAK-1m

A different submarine badge is awarded to Special Operations personnel who qualify for midget submarines. This badge is a similar to the PAK-1 & 2 but with a more angular submarine profile and an upward-facing short sword in the center of the wreath and the sail (PAK-3). It is issued in gold for all qualifying personnel, officers and enlisted, though, according to sources, silver is also available. Fake versions exist with a similar design but completely flat that are issued in gold, silver, red and black.

PAK-3

For a time, cadets in the Pakistani Naval Academy who completed a submarine orientation course, including spending time onboard a submarine, earned a gold cadet badge that was basically the same as the officer's badge, but without the wreath (PAK-4). They could continue to wear the badge until they had graduated from the Naval Academy. This was offered as a means to attract cadets into the submarine service. The effort did not attract volunteers as hoped, so the badges are now obsolete.

There was a similar, but shorter-lived effort for engaging enlisted personnel. Very briefly, silver "cadet" badges were provided to enlisted personnel completing submarine training on the PNS *Himalaya* (PAK-5). These badges were the same as the PAK-4, but in silver, and they were quickly discontinued.

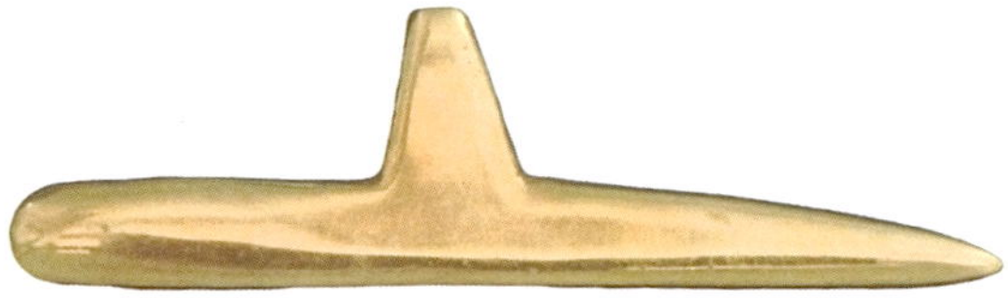

PAK-4

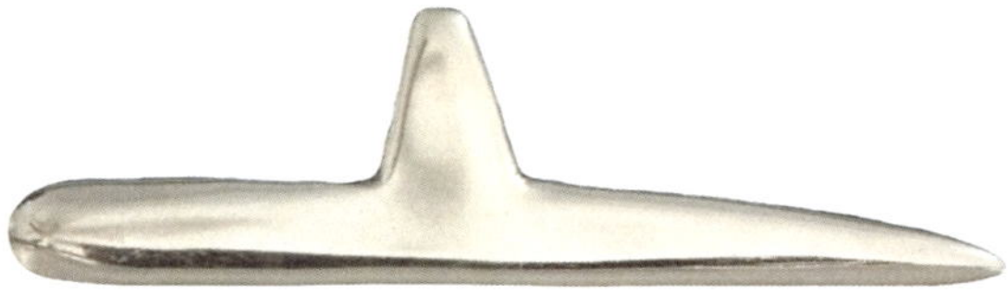

PAK-5

Officer and Enlisted bullion patches were used for a period of time with blue sea-going uniforms (PAK-1C & PAK-2C). These patches eventually became obsolete due to the qualification badge no longer being authorized for wear on the sea-going uniforms.

PAK-1C

PAK-2C

PERU (PER)

Peru was on the forefront of submarine development and implementation with a successful submarine being tested as early as 1880. Thirty years later, now senior Naval leadership remembered the success of that time and convinced the government to order two submarines from France in 1910. The Peruvian submarine service was established when those first two boats were received at the French shipyard on 19 August 1911, making it the oldest submarine service in South America. There is no known record of any Peruvian submarine distinctive during these early days.

Between 1926 and 1928, four *R*-Class (*Erres*-class) boats were purchased by Peru from the Electric Boat Company in the United States forming the country's first official Submarine Division. At the time of purchase, Peru negotiated a deal where the U.S. would also qualify the Peruvian officers on the submarines. So, it made sense that when the Peruvian Naval Command issued official uniform requirements for the submarine force in July 1927, they included a description for a gold qualification badge for officers (PER-1*) that was basically a duplicate of the U.S. officer's qualification badge. This badge was worn on the left chest, above the ribbons.

Initially, enlisted personnel had no qualification insignia, but that changed in 1932 when they were authorized to wear the same design as U.S. enlisted personnel, only in gold thread on a white cloth background (PER-2*). This was worn on the right sleeve, between the shoulder and the elbow. The authors have been unable to find valid original examples of either of these badges to show.

While personnel assigned to submarines had wanted a Peruvian-designed insignia since the formation of the submarine force, it was not until 1942 that the effort gained some momentum. At that time Ensign Jorge del Rio Roldan, a surface sailor who was temporarily assigned to the *R-3*, drafted a suggestion for a new Peruvian submarine service badge and presented it to the boat's captain. This insignia was the Peruvian national coat of arms on a Naval anchor, superimposed on a surfaced submarine and a torpedo, with two diving heraldic dolphins underneath and a scroll above with "*Division De Submarinos*" on it (PER-ILL-1).

PER-ILL-1

According to sources, it is possible a prototype of this badge was made by U.S. jeweler Bailey, Banks & Biddle, but they made a mistake when creating the national coat of arms. While the effort was appreciated, the Naval Command felt that the badge had too many components, and it was returned with "observations" by the end of 1943. However, this initial effort prompted, now Second Lieutenant del Rio, to improve upon it and offer another six designs.

Working with an artist from the National Mint, they finally arrived at a design that incorporated a profile of a port-facing *Erres*-class boat with the national coat of arms in the center, supported by two inward-facing heraldic dolphins representing the ability to perform at depth with intelligence and skill. This badge was officially approved by Supreme Resolution No. 1186 on 8 June 1949 and was issued in gilt (gold) for officers (PER-3) and silver for enlisted personnel (PER-4). The new badges were first worn when all four *Erres*-class submarines went back to Electric Boat in 1950 for refit and modifications. This remains the current design of the Peruvian submarine qualification insignia.

PER-3

PER-4

The early badges were typically a solid back with horizontal pin attachment and worn on the left chest above the ribbon bar if worn. At some point, possibly in the 1960s, these transitioned to being exclusively hollow back with clutch pin attachments. Miniature versions of these badges are approved for use on mess dress uniforms that are also hollow back (PER-3m & PER-4m).

PER-3m

PER-4m

Because of their presence in photos, patches appear to have been recently approved for use on submarine coveralls by only officers, with dark blue or black rectangular patches with gold thread submarine insignias known (PER-5C.1* & 5C.2*), but the authors have no examples to show.

There are multiple "Astosubpe" badges and patches that have been issued by the Peruvian submarine veteran's association. While quite ornate and colorful, these are veteran badges only and have no official standing with the Peruvian military.

POLAND (POL)

The Polish submarine service was established in 1931 with the commissioning of three *Wilk*-class mine-laying submarines purchased from France in 1929. In the late 1930s two Dutch-built (with Dutch/Polish engineering collaboration) *Orzel*-class boat were added to the fleet. During this period, there were multiple pins and other items made for several of these boats, mostly as a means for raising public funds to support their construction, but initially there were no specific submarine distinctives.

This changed with the commissioning of the ORP *Orzel* (Eagle) in 1939 ("ORP" standing for *Okret Rzeczypospolitej Polskiej*, or "Ship of the Republic of Poland"). As the lead ship of the new class of submarines being built for the Polish navy, there was a strong sense of national pride linked to the *Orzel*, and a significant portion of the funding for the construction was obtained fund raising from the Polish population. Reflecting this pride, the commissioning crew of the *Orzel* was presented with a distinctive identifying them as the "first crew". This badge was in the shape of a shield, silver in color, with anchor arms and flukes framing the bottom, topped by the anchor stock and ring, with a starboard-facing profile of the submarine in the middle on waves, "*PIERWSZEJ ZAŁODZE*" (First Crew) "ORP "ORZEL"" above the submarine and "1939 LMK" below it (POL-ORZ). This badge is incredibly rare since the majority of the badges went down with the crew when the *Orzel* was lost in May of 1940. Two known examples exist in the Polish Naval Museum in Gdynia and in the British National Maritime Museum in Greenwich. While possibly technically a commemorative badge, we include it here due to its importance in Polish submarine history.

POL-ORZ

Though Poland continued to operate submarines through the 1950s, it was not until the early 1960s that the Polish navy began to consider a distinctive for submariners. Indonesian submariners in Poland for training had developed a submarine badge, which started Polish submariners considering the possibility, however, the fact that all of their neighboring countries had a submarine distinctive and Poland was still without one is what eventually turned the idea into action. Designs were generated by personnel within the submarine squadron and these ideas were provided to an artisan, Mariusz Ditberner, who refined the designs. His proposal was submitted to the Ministry of Defense (MOD), and in 1964 the official Polish submariner qualification insignias were approved by Polish MOD Order #9.

This first qualification badge was a port-facing diesel submarine featuring prominent saddle tanks, drive shaft, prop, rudder and periscope, a stylized Polish white eagle embossed in the middle astride the arms and flukes of an anchor. This badge was issued in three classes: Gold for those qualified to command a submarine (not just Commanding Officers) (POL-1.1), silver for other submarine qualified ship's officers after being aboard one year, and for enlisted ratings aboard a submarine for greater than 10 years (POL-2.1), and bronze for enlisted ratings personnel earned after 5 years aboard a submarine (POL-3.1). This is, by far, the longest period required for enlisted personnel to earn a submarine badge of all of the countries researched for this book. These badges are worn by all personnel on the right chest on all uniforms.

POL-1.1

POL-2.1

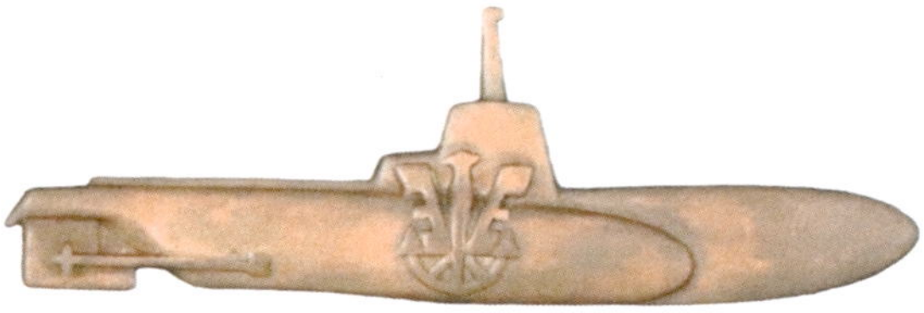

POL-3.1

According to sources, approximately 200 of Mr. Ditberner's original die pressed badges were issued. This first version was hollow back with a single center threaded fastener. A few years after the initial badge distribution the mint created a higher-quality badge with a solid back, but the gold variant of this version had problems with the gold-colored clear coating wearing off and exposing the silver base metal (POL-1.2, POL-2.2, POL-3.2). A few more variants were issued over the years by the government mint, all hollow back/cliché with assorted minor differences to the eagle or the submarine profile. There were never any miniature badges created for the 1964 design badges.

POL-1.2

POL-2.2

POL-3.2

In 1990, following the fall of communism in Eastern Europe, the white eagle on the Republic of Poland's coat of arms had its crown restored that had been removed by the communist government of the Polish People's Republic in 1955. Soon after, most military and government symbols and awards that incorporated the eagle were modified to reimplement the crown. However, the submarine insignia was not one that saw the crown restored. In the late 1990s a drawing for a newly proposed badge using the Polish eagle with the crown was introduced, but never gained any traction with squadron. Bypassing the approval process, a badge manufacturer named Adam Lis created a few batches of badges based on this drawing.

This new design took on a more sleek and stylized appearance, discarding the drive shaft and prop of the 1964 design. Still a profile of a port-facing submarine, a finely detailed crowned Polish white eagle more representative of the national coat of arms is in the middle atop the arms and flukes of a fouled naval anchor. Within a few years, Stanislaw Sliwa of the badge manufacturer Celstan improved upon the design and badge quality and issued the badges in larger numbers. As with the previous versions, this badge is also issued in gold (POL-UA1.1), silver (POL-UA2.1), and bronze (POL-UA3.1). For this design, the eagle and anchor are the same color as the badge. Miniature versions of these badges were used by personnel on their working uniforms and coveralls (POL-UA1.1m, POL-UA2.1m, POL-UA3.1m). At first, the full-size badges were not widely used as the 1964 design was still the only approved badge. But slowly, they were more commonly used and tacitly accepted by authorities to be worn in place of the 1964 badge design without ever receiving official authorization.

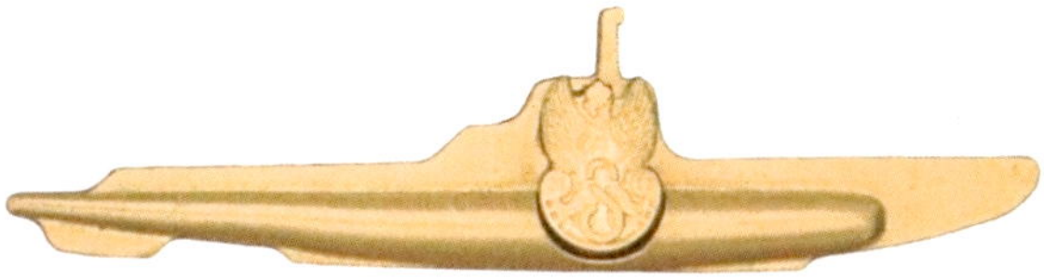

POL-UA1.1

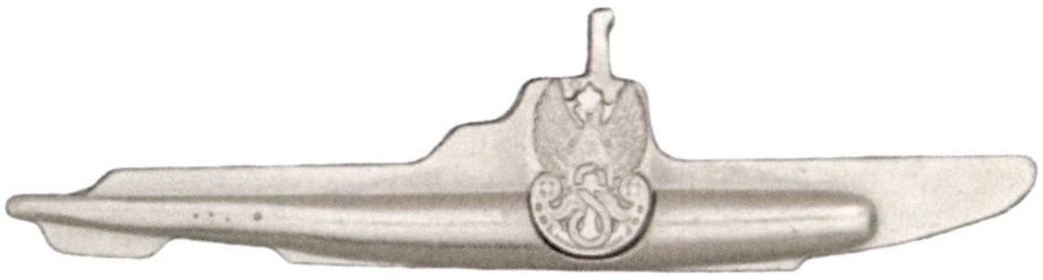

POL-UA2.1

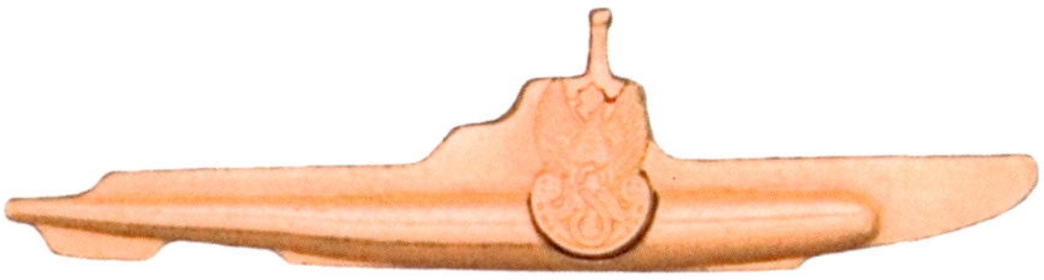

POL-UA3.1

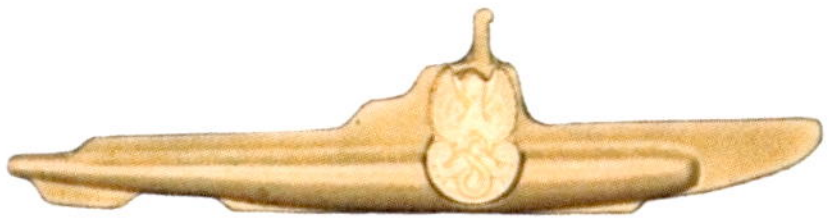

POL-UA1.1m

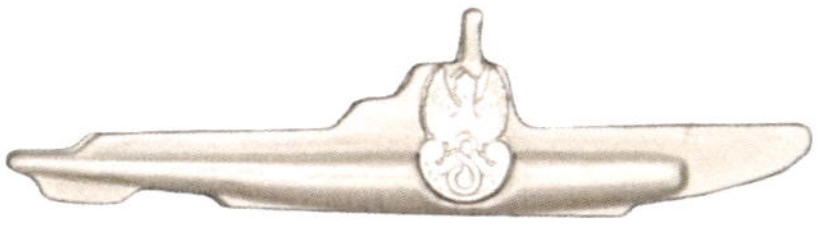

POL-UA2.1m

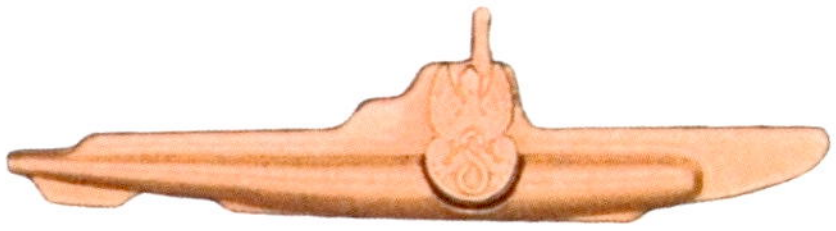

POL-UA3.1m

In the mid-2000s several new designs appeared, including some that were intended to be submitted as new official badges. One was made of hammered/beaten metal and used the old Ditberner design but included an embossed modified Polish eagle with a crown. Because of the lack of detail in the embossment, the crown looked more like spikey hair, and, in general, the badge was disliked (POL-CON-1). Without official sponsorship, its status was relegated to that of just a concept badge. Another had a profile similar to the 1964 design, but thicker, with an embossed Polish white eagle with crown (POL-CON-2). A third featured a port-facing *Kobben*-class submarine profile with a silver Polish white eagle in the center (POL-CON-3). Like all other badges, these were available in gold, silver, and bronze and miniature versions were also produced. These badges are shown as a warning to collectors. These and a few other "concept" badges have been produced in relatively large numbers by manufacturers trying to take advantage of the collector's market for Polish submarine badges, and are not officially sanctioned badges, nor used by the submariners.

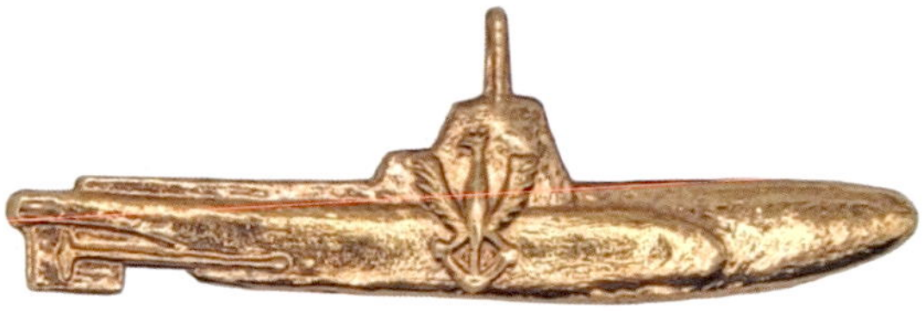

POL-CON-1

POL-CON-2

POL-CON-3

It was not until 2014 a new badge was actually authorized for Polish submariners. On 8 April 2014, the Ministry of Defense issued Decision No. 122 approving a new design created by a well-known Warsaw manufacturer, Piotr Olek. This new badge paid homage to the original design by utilizing a similar submarine profile with the pronounced saddle tanks, drive train and periscope, but implementing a new significantly more detailed Polish naval eagle with a tall crown. This badge was also issued in three classes, with the naval eagle being a pewter gray for all three (POL-4, POL-5, POL-6). Miniature badges were approved in the regulation for wear on the lapel of civilian clothing (POL-4m, POL-5m, POL-6m).

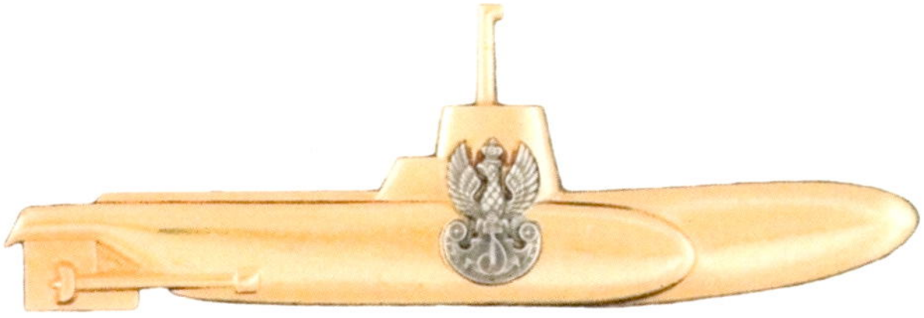

POL-4

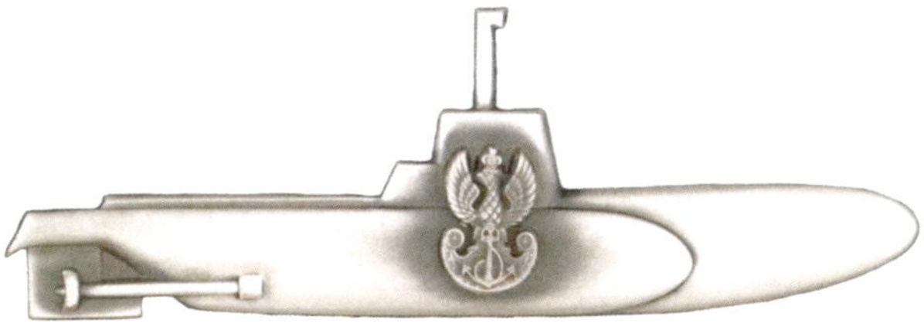

POL-5

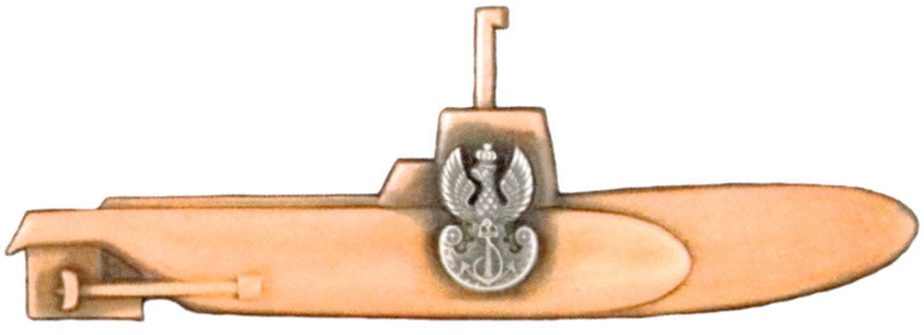

POL-6

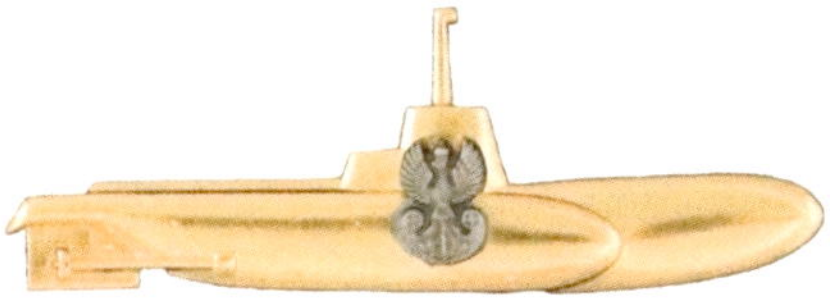

POL-4m

POL-5m

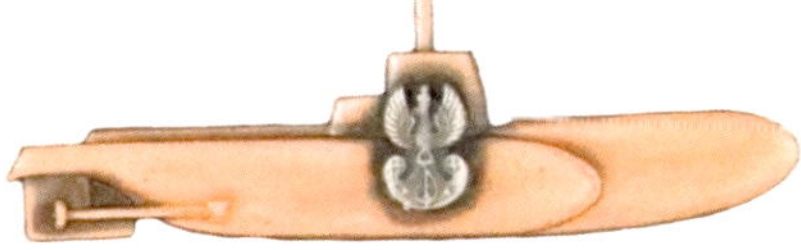

POL-6m

Shortly after the Olek badges were issued, a different manufacturer made badges with the same design as the Adam Lis badges, but using the eagle from the Olek badges in silver on all versions (POL-UA1.2, POL-UA2.2, POL-UA3.2). Miniature versions of these badges also exist (POL-UA1.2m*, POL-UA2.2m*, POL-UA3.2m*). Again, these badges have no official standing within the navy, however, are used by Polish submariners.

POL-UA1.2

POL-UA2.2

POL-UA3.2

When Poland purchased a *Kilo*-class submarine from the USSR and commissioned it in 1986 as the ORP *Orzeł III*, a commemorative badge designed and manufactured by the Warsaw manufacturer Zygmunt Olszewski was issued for the crew members reminiscent of the first *Orzeł* badge. This badge was a silver oval wreath surmounted by a white Polish eagle on a red shield, the words, "*PIERWSZA ZAŁOGA* 1986" (First Crew 1986) on the bottom circumference, and in the middle a stylized, slightly starboard-facing, black *Kilo*-class submarine profile with the words "ORP ORZEŁ III" below it on a red enamel sea (POL-ORZ-III). Again, while technically a commemorative badge, this was worn proudly by the crew on their uniforms. Multiple restrikes and fakes of this badge exist.

POL-ORZ-III

The designer proposed several other variations of this badge, including an all-silver version with a blue lower circumference of the wreath for conscripts, and two gold versions for VIPs. While none of these options were ever officially approved and therefore have no standing in the Polish navy, they were still manufactured for the collecting community (POL-ORZ-UA1.1, UA1.2, & UA1.3), and even fake versions of these unauthorized badges exist.

POL-ORZ-UA1.1

POL-ORZ-UA1.2

POL-ORZ-UA1.3

In the 1990s, the Polish submarine force began implementing boat badges like many other countries have done. But unlike many countries, Polish submariners actually earn their boat badges through an approximately six-month long process involving passing a board with the division officer, culminating with a ceremony where the boat badge is presented, and can be officially worn on the left chest of their uniform. These badges were designed by Piotr Olek and were a fouled naval anchor having the name of the boat arched over the top and the head of the boat's namesake mounted in the middle. Badges were created first for the *Orzel* (Eagle) and *Dzik* (Boar), and then a few years later for the *Wilk* (Wolf). Following the purchase of four *Kobben*-class submarines in 2002-2003, badges were issued for the *Kondor* (Condor), *Bielik* (White-Tailed Eagle), *Sokol* (Falcon), and *Sep* (Vulture). These original versions can be difficult to find, and fakes exist. Small badges that are just the animal heads, known by the Polish submariners as "diamonds", were used as

Orzel (Eagle)

Dzik (Boar)

Wilk (Wolf)

Kondor (Condor)

Bielik (White-Tailed Eagle)

Sokol (Falcon)

Sep (Vulture)

replacement badges. There have been multiple unauthorized boat badges created solely for the collector's market that include submarine profiles, with the boat's name or animal heads on them. None of these are used by the Polish navy.

A Polish submarine squadron badge was proposed in 1995. Squadron personnel took the initiative to design, approve, and gather the funds to fabricate this badge. Working with Piotr Olek, three prototypes were created, from which one was approved and implemented in 1996. This badge is a silver Polish eagle sitting astride a port-facing *Wilk*-class (the first Polish submarine) submarine profile in front of a blue enamel Maltese cross, with four golden anchors as the back of the badge, their arms forming the corners. On the arms of the cross are the words, "*DYWIZJON OKRĘTÓW PONWODNYCH*" for "Submarine Squadron" and the date "30 IV 1932", which was the date the squadron was established (POL-SQD-1.1). The first batch of these badges had the manufacturer's initials. "PO" stamped on the reverse. Later versions have a different pentagonal stamp with the maker's mark of a mermaid with a sword and shield, the year "1977" and "P.OLEK". A version in all silver/pewter also exists (POL-SQD-1.2) that was created as part of the prototype process, so very few exist. Multiple fakes of these early versions have been manufactured. In 2017 the squadron slightly redesigned the badge, with the new version being similar to POL-SQD-1, but with slightly darker blue enamel on the cross, reducing the arc at the ends of the cross arms, and removing the words (POL-SQD-1.3).

POL-SQD-1.1

POL-SQD-1.2

POL-SQD-1.3

With the Polish government currently pursuing replacement submarine options for the retired *Kobben*-class submarines, there is a very real possibility we may see new Polish badges in the next several years.

PORTUGAL (POR)

The Portuguese submarine force was born on 15 April 1913 when the Portuguese navy accepted the NRP *Espadarte* at the shipyard in La Spezia, Italy. While having only a single submarine until 1916, the Portuguese Navy quickly adopted a submarine insignia.

Authorized 21 August 1915 under ordinance #455, the first Portuguese submarine distinctive depicted a starboard-facing *Espadarte*-class submarine profile. Worn on the right sleeve five centimeters below the elbow, this badge was embroidered in gold thread for officers and senior enlisted personnel (POR-1), and in red thread on dark blue backing (POR-2.1) for junior ratings' winter blue uniforms. Sometime after the badge was approved, a blue on white (POR-2.2) badge was authorized for junior ratings' summer whites. Rumors have persisted for years regarding metal versions of the early Portuguese badges made before 1960, however the authors have been unable to uncover any evidence to support the existence of such badges through official channels, though it is possible jeweler-made versions exist.

POR-1

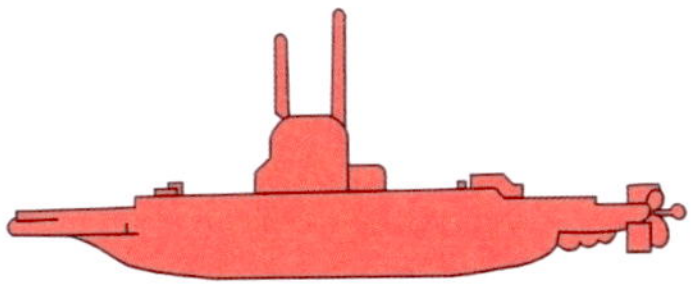

POR-2.1

POR-2.2

By 1920 the Portuguese Navy had a few more submarines, and a new submarine badge was implemented on September 7th of that year per ordinance #2:424. The new badge reflected their newer submarines and was a port-facing submarine profile, this time with a detached representation of the national coat of arms placed above the submarine conning tower. The ordinance approved a badge "embroidered in gold relief on a black background" (POR-3.1) for all personnel and worn on the upper right chest of all uniforms. In 1925, the junior ratings badges kept the same profile with the coat of arms above but changed to red and blue embroidered patches on the right sleeve (POR-3.2 & POR-3.3).

POR-3.1

POR-3.2

POR-3.3

In 1936, changes were made that removed the national coat of arms from above the badge for all enlisted personnel, but the submarine profile still remained the same (POR-4.1, POR-4.2, & POR-4.3). Because the authors have found no valid examples of any of these early Portuguese badges except for the POR-3.1, we are utilizing sketches provided by the Portuguese submarine squadron as examples of the Portuguese submarine badge evolution.

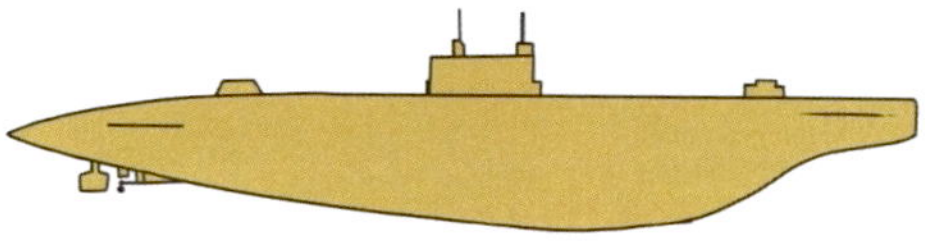

POR-4.1

POR-4.2

POR-4.3

The POR-3.1 for officers and the POR-4s for enlisted personnel remained the Portuguese submarine insignia until Government Journal No. 213 was issued on 16 September 1959 describing a new badge that was implemented in 1960. This new badge design was a stylized *Narval*-class (British *S*-class) submarine, again a port-facing profile, but this time more rounded and with pronounced saddle tanks and an angular conning tower. For officers, this badge was in gilt metal with a gilt national coat of arms attached above the periscopes (POR-5 (Sketch)) and Senior enlisted/petty officers had the same badge but without the national arms (POR-6). Junior ratings used the same design, but in the red (POR-7.1) and blue (POR-7.2) patches. Versions of both patches are known with port and starboard-facing submarine profiles. The officer's badge was changed in 1961 to

POR-5

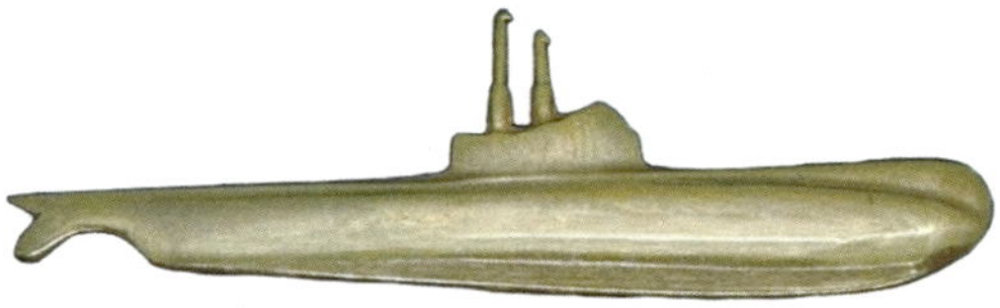

POR-6

POR-7.1

POR-7.2

incorporate a high-quality cloisonne national coat of arms in the middle of the submarine profile (POR-8). Both of these badges are solid back, with a horizontal pin attachment, or with two clutch pins. Bullion versions of the POR-6 (POR-6C) and the POR-8 (POR-8C) have been issued.

POR-8

POR-6C

POR-8C

Decree No. 21 688 issued to the uniform regulations on 26 November 1965 changed the submarine insignia to a slightly new design that was 25mm tall, 65mm long, with the officer's badge having a slightly less detailed 9mm diameter national coat of arms and laurel wreath in the center. This new design had the same basic profile and was the same length as those originally issued, but with a thicker hull that has a slight upward curve, a slightly more rounded tail and a more pronounced V-shaped section between the hull and rudder (POR-9.1, POR-9.2 & POR-10). These badges are a thinner metal with hollow backs and a horizontal pin attachment and remain the approved designs for the officer and petty officer Portuguese submarine badges.

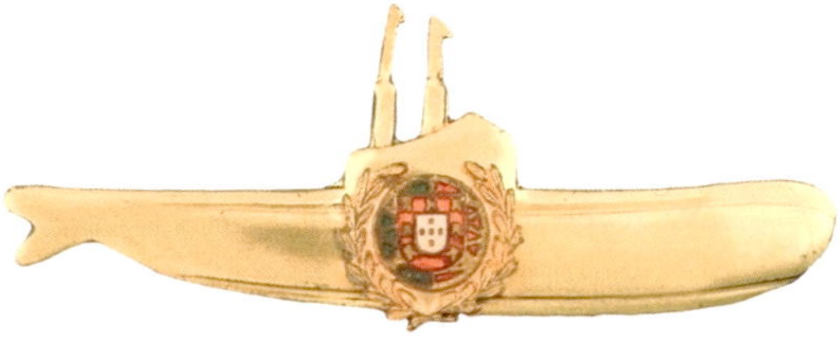

POR-9.1

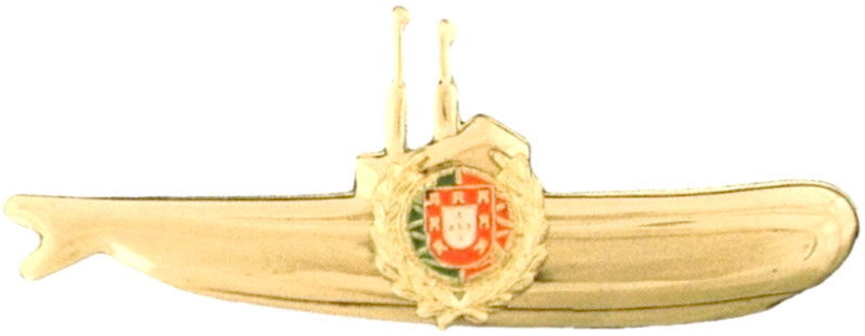

POR-9.2

POR-10

At some point a metal badge with a completely different profile was produced. Sources estimate this badge appeared from the 1950s to the early 1960s and was possibly one of the options being considered for the badge design change in 1960. This gilt badge is a much more detailed British *S*-class submarine port-facing profile with a more angular bow and a tall extended periscope. Issued in an officer's version with an integrated national coat of arms (POR-UA-O) and a senior enlisted/petty officer's version without the national arms (POR-UA-PO), these are solid back with a horizontal pin attachment method. No known cloth versions have yet been identified, and the source of these badges is unknown, however, they were never approved for use by regulation, and if used, were only used in an unofficial or trial capacity.

POR-UA-O

POR-UA-PO

ROMANIA (ROM)

To ramp up their military capabilities in the Black Sea, the Kingdom of Romania requisitioned a submarine from Italy in 1927. This submarine was not completed until 1931 and was not accepted by the Romanian Navy until 15 August 1936 when the NMS *Delfinul* was commissioned. *Delfinul* was the only full-size Romanian submarine in service until WWII, finally being joined by two others, the *Rechinul* and the *Marsuinul*, in May of 1941.

A submarine badge was created for the original *Delfinul* crew, implemented around the time of, or slightly before, its commissioning. This silver badge features a port-facing leaping shark on a ring with laurel wreath boughs tied at the bottom and the word "SVBMARIN" across the top arc with the royal crown surmounting the wreath (ROM-1). The dorsal fin of the shark separates the "B" and the "M", making it appear as two words. The badge is very reminiscent of the Italian officer and petty officer badge (ITL-2 & ITL-3). Examples are known with both pierced and solid crowns and was worn on the left chest above the pocket. According to sources, this badge was awarded to all personnel aboard the *Delfinul*. There are known copies/ reproductions of this badge in both silver and gold that are obviously not valid badges.

ROM-1

On 19 April 1943, the next Romanian submarine insignia was authorized. Known as the "Romanian War Badge", it was very similar in design to the ROM-1, with the only differences being the word "SUBMARINE" now across the top arc of the ring and the shark's dorsal fin no longer bifurcating the "B" and the "M". These regulations specifically identify the figure in the middle should be that of a dolphin, but Romanian military historians disagree, noting that the prominent gills on the previous badge and the pointed nose and presence of a second dorsal fin identify it as a shark. It was issued in three versions, "gilded" for officers, "silvered" for senior ratings, and "nickel-plated" for junior enlisted personnel (ROM-2.1, ROM-2.2, ROM-2.3) according to the regulations. These badges were issued to crew members of all three Romanian submarines. Plaques with the words, "*Audaces Fortuna Juvat*" (Fortune Favors the Bold), the motto of the Romanian submarine force, were mounted on the forward part of the sail of all three submarines.

ROM-2.1

ROM-2.2

ROM-2.3

In 1944, all three submarines were confiscated by the USSR, with *Delfinul* and *Rechinul* not returned until 1951, *Marsuinul* having been sunk by an explosion of its own torpedo while under Soviet control in 1945. *Delfinul* remained in service until 1957, and *Rechinul* until 1961, both eventually being scrapped by 1967. It is not known if badges were issued to submariners assigned to the submarines during that period.

Romania's next foray into a submarine service occurred when a *Kilo*-class submarine was purchased from the Soviet Union and commissioned in August 1985. Also named *Delfinul*, it was only in service for 10 years before being shifted to inactive status in 1995 due to lack of batteries. Despite being relegated to inactive status, the submarine has maintained a crew since 1995.

A badge was issued for these submariners in the late 1980s/early 1990s. This badge is a port-facing submarine profile with a shield in the center featuring the red, yellow and blue of the Romanian flag. According to sources, a trial run of the badge was made in gilt, silver, and copper/bronze to see how it would look in different metals, however, only the gilt badge (ROM-3) was put into production and issued to crew members. Collectors should be aware that fake Romanian badges of all three metals have been manufactured in large quantities. While *Delfinul* was operating as an active sub, the badge was issued to all submariners following completion of a six-month submarine course. After the submarine became inactive, the badge became a designation badge, not a qualification badge, and was issued to all personnel assigned to the submarine.

ROM-3

There is a badge featuring an eagle with crossed swords shown in other books as a possible Romanian submarine badge from the 1990s. Sources within the Romanian Navy have verified the badge is not, and has never been associated with their submarine service, and is likely of Slavic origin.

Though Romania has not actively operated submarines since the batteries died on the *Delfinul*, as of this writing, they have recently approved a large procurement of naval equipment, including submarines. Whether or not that allotment results in new Romanian submarine insignia, we shall soon see.

RUSSIA / SOVIET UNION (RUS)

Imperial Russia experimented with various submarines as early as the mid-1800s, but the navy did not commission their first submarine until the *Delfin* (Dolphin) was commissioned in 1903. They quickly ramped up submarine purchases, signing an agreement with Simon Lake in the U.S. for seven *Osetr*-class submarines commissioned in 1905, and with Germany for three Type-*E* submarines commissioned in 1907. Eager to have an operational submarine for use in the Russo-Japanese war, they purchased a German-built submarine that was commissioned in 1904 as the *Forel* (Trout) and immediately sent by rail to the Vladivostok. Though *Forel* never saw combat, its mere presence supposedly had a significant psychological impact on Japanese forces, solidifying Russia's belief in the value of submarines. While rumors persist that a metal badge or two may have been commissioned to commemorate it, no conclusive evidence of these badges exists.

Though the *Delfin* had been in use since 1903, the Imperial Russian navy did not classify submarines as an independent class of naval ship until 1906. When this happened, the Naval Command quickly convened a submarine officer navigation class, with the first class graduating in 1907. On 26 January 1909, by Imperial Decree #31454, Imperial Russia became the first country in the world to officially implement a <u>metal</u> submarine distinctive when a special badge was issued to graduates of the submarine officer training class as qualified "Submarine Officers". The badge was made for the recipients by various jewelers in the St. Petersburg area based upon an issued drawing. The badge is made of silver, with a stylized *Delfin*-class submarine resting atop the flukes of a fouled Hall's anchor, a fine rope weaving through the ring and around the shank and arms, all completely encircled by an anchor chain (RUS-1). It is screw back with a plate, or "nut" securing it to the uniform. Importantly, on most genuine examples of this badge, the nut and the back of the badge itself have St. Petersburg jewelers' hallmarks. That said, an example is known with excellent provenance, but no jeweler's hallmarks. Multiple fake versions of this badge exist.

RUS-1.1

RUS-1.2

RUS-1.3

Despite having an approved drawing, because the RUS-1 badge was made by several different jewelers there are multiple variations of this badge with differences in the stockless anchor details, number of chain links, and the style and weave pattern of the rope (RUS-1.1, RUS-1.2). The most consistent aspect of the multiple variations appears to be the general body style of the submarine. A badge was issued for the graduating class of 1913 that was also jeweler-made but was different in that instead of the stockless Hall's anchor, it featured an admiralty anchor with a full stock and much wider and thinner arms (RUS-1.3). The RUS-1 badge was worn on the right chest of the qualified submarine officer. This badge, along with all other military distinctives, was banished following the "October Revolution" in 1917 when the Russian Communist Party overthrew the monarchy and established the Union of Soviet Socialist Republic (USSR), also known as the Soviet Union.

According to sources, it is possible that graduates of the school had their own badges made for post-1917 graduations. These sources also report the school began reissuing badges in 1927, and that a description of this badge exists, but this information cannot be confirmed by the authors.

Of the graduates in the initial submarine navigation class, there was a single Bulgarian. Lt. Rashko Serafimov qualified to earn the Submarine Officer badge with his classmates upon graduation in 1907, however, as the RUS-1 badge was not approved until 1909, no one in the class received the badge until well after graduation. It is not fully understood how Lt. Serafimov would have received his badge, but in 1969, when his heirs donated his qualification badge to the Bulgarian Navy Museum in Varna, the badge raised more questions than it answered. The so-called "Serafimov badge" is similar to the RUS-1 badges presented to his classmates being made of silver with a *Delfin*-class submarine in the center and backed by a circular anchor chain. However, it is different in that the outer section is a wreath of oak and olive branches, topped with what could possibly be an Imperial Russian double-headed eagle coat of arms over two crossed anchors (RUS-SER). While many theories exist, it has not been conclusively identified why this badge is so different from the others, nor exactly when the badge was presented to him. It is possible that this badge may have been presented to a small number of Bulgarian submariners as a few very similar badges have been identified.

RUS-SER

It was not until 21 May 1942 with the outbreak of WWII that badges were allowed to be used again within the USSR military. For submarine personnel, badges to honor "Excellent Submariner" and "Excellent Torpedo Man" within the enlisted ranks were created. As with the RUS-1, these badges were made by multiple manufacturers, resulting in design and measurement differences. In general, the badges are an ornate bronze shield flanked by laurels of oak leaves, a bronze Soviet hammer and sickle on red enamel in the center, surrounded by "Excellent Submariner" or "Excellent Torpedoman" on a circle of white, with a bronze submarine profile below for the submariner badge (RUS-2.1 – 2.4), and a bronze torpedo and a mine for the torpedoman badge (RUS-3.1 – 3.3). The multiple manufacturers, including MTX, ШМЗ/НКПС (SHMZ/NKPS), and 3-д «Победа» (3D "Victory") (as inscribed on the backing nut), resulted in variations of badge sizes, shield designs, submarine profiles used, and the length of the attaching screw with at least four different styles of each existing. These badges were rarely issued following the war and were eventually discontinued. Multiple excellent modern replicas of these badges exist, making it difficult to differentiate between originals and copies.

RUS-2.1

RUS-2.2

RUS-2.3

RUS-2.4

RUS-3.1

RUS-3.2

RUS-3.3

On June 12th of the same year, Order #170 of the People's Commissar of the USSR Navy authorized the establishment of a "Submarine Commander" badge. Compared to the previous badges, this one was quite simple showing a port-facing gilded bronze *Shchuka*-class detailed submarine profile, with a bronze hammer and sickle on an enamel red star in the center (RUS-4). This was screw back and worn on the right chest. At 70mm long and no more than 17mm high, it was only in use for a little over a year making genuine versions of this badge very difficult to find.

RUS-4

On 16 October 1943, the People's Commissar of the USSR Navy Order #383 was issued, changing the previous Submarine Commander badge. Most notably, the hammer and sickle were removed from the center of the red enamel star. Additionally, the size was modified to be 67mm long and 22.5mm high (RUS-5). Though there were small variations in size and details used over the years, the basic design of this badge was in use for almost 50 years. Shown here are: One of the first issued badges from 1943 (RUS-5.1), and the second issue badge from 1943 with a painted star (RUS-5.2).

RUS-5.1

RUS-5.2

A late 1940s variation with a red enameled star (RUS-5.3), and an example from the 1970s-1980s (RUS-5.4).

RUS-5.3

RUS-5.4

An example from the late 1980s-early 1990s (RUS-5.5), and an aluminum example from that same time period (RUS-5.6).

RUS-5.5

RUS-5.6

On 22 May 1961, the Soviet Navy leadership authorized the presentation of an extended mission or "Distant Cruise" badge to submarine crews when the Commander-in-Chief of the Soviet navy issued order N 149. This badge had a profile of a submarine on a field of light blue, sailing below the Soviet Naval ensign, with the words, "For a long voyage" in Cyrillic on a red banner at the bottom (RUS-6). This badge had multiple variations with different submarine profiles, including one issued in the late 1960s for Operation "Ocean" that was issued with a separate scroll hanging from the bottom of the badge with the name of the operation (RUS-6-OCEAN). This badge was used until 1976 when it was replaced by a new style badge under Order No. 2 of the Minster of Defense on 2 January 1976.

RUS-6

RUS-6-OCEAN

The new style Long Voyage badge was a bronze pentagonal shield with the profiles of more modern submarines in the center, a compass rose above it, and the Soviet Naval ensign at the top. Under the submarine the words, "For a long voyage" etched on a bronze bar and a naval anchor in a shield on a striped background underneath (RUS-7). As with the previous version, multiple designs with various submarine profiles exist. On 15 August 1987, the Minister of Defense issued order No. 233, abolishing the 1976 version and reimplementing the 1961 model (RUS-6). Again, multiple different badges appeared with different submarine profiles and slight badge differences due to different manufacturers.

RUS-7

In 1992, following the fall of the Soviet Union, new military badges were issued by the Russian Republic Navy. The new Submarine Commander badge maintained the profile of the previous version, but the red star was removed (RUS-8). The Long Voyage badge retained the same basic formats of the RUS-6 and continued to use various submarine profiles but replaced the Soviet Naval ensign with the Andreevsky Russian Naval ensign (RUS-9). These are the last known official submarine badges issued by the Russian Navy.

RUS-8

RUS-9

During the mid-1990s there were efforts by different factions within the submarine force to develop new badges. One effort featured gold Submarine Commander and silver officer badges, placing the Russian double-headed eagle in the center of a port-facing submarine profile (RUS-UA1). Another featured a submarine profile with a globe in the center, laurel boughs underneath, and a number center bottom that was intended to represent the number of combat missions conducted by the commander (RUS-UA2). While none of these efforts resulted in any officially authorized badges, examples are shown for reference. Though some of these badges were manufactured, they were never officially sanctioned nor adopted. There was also an attempt to implement use of a badge similar to the 1909 RUS-1 badge, but again, this had no official support. However, Russian officers have been seen with the unauthorized RUS-1 copy officer badge on their uniforms.

RUS-UA1

RUS-UA2

At this point it is necessary to address the elephant in the room. Beginning in the 1950s, unofficial commemorative, memorial, and representative pins/badges started to appear in the Soviet Union. The numbers of these significantly increased during the 1970s and 1980s and continue to be fabricated to this day. There are easily close to a thousand different designs and variations, making the Soviet Union/Russia the most prolific producer of submarine-related insignia, even if the vast majority are not authorized for uniform wear. Examples include *Dembel*, or Demobilization badges given to submariners exiting the service, with the *Komsomol*, or Communist Youth League logo on it, commemorative badges for multiple different Soviet/Russian submarine Projects, and dozens of others. An excellent book on this topic, *Badges and Medals of Russian Submarines* by A.A. Dmitriev was published in 2002 and is a must-have resource for those interested in these insignia, but even this book only scratches the surface of the manufactured pins/badges/insignia. There are several other reference documents for Russian submarine-related materials as well.

SINGAPORE (SIN)

Singapore's submarine force was christened in 1997 with four *Sjöormen*-class submarines purchased from Sweden. Around that same time the Republic of Singapore Navy (RSN) submarine insignia was introduced.

According to sources, the RSN submarine badge was designed by the wife of one of the first submarine captains. The badge is gilt with two bottle-nosed dolphins leaping out of the water away from the center of the badge, which features the RSN crest of a fouled naval anchor on a wreath of oak boughs and the Singapore national coat of arms in the center. The badge issued to all qualified submariners is gilt, with officers and enlisted personnel wearing the same badge on the left upper chest above the ribbon bar (SIN-1). The initial badge presented to a submariner is a serialized solid-back badge, and any subsequent badges purchased by the submariner are typically hollow back versions from the base exchange. Many submariners wear the hollow back versions during parades due to the badges being lighter and not sagging as much during wear.

SIN-1

No miniature versions of the metal badge are used. For dress white uniforms a full-sized bullion badge is used (SIN-1C). There are several versions of this badge due to different manufacturers, with some having a double screw post attachment method. A miniature of this badge is used for the mess dress uniform that is available with bullion (SIN-1Cm.1) and thread (SIN-1Cm.2) versions.

SIN-1C

SIN-1Cm.1

SIN-1Cm.2

There is a silver version of the metal badge that is issued to "Honorary Submariners", typically VIPs within the military being recognized for their contribution to the submarine force (SIN-VIP). These are extremely rare with fewer than 20 valid badges being issued. Fake versions exist in fairly large quantities

SIN-VIP

The RSN submariners are relatively unique in that they are currently authorized to wear qualification insignia from three other countries on their uniforms. They are allowed to wear submarine badges from Sweden, the Netherlands, or Germany. This is due to personnel qualifying on the submarines in these countries. Very few personnel qualify on the Dutch or German submarines, and those small numbers are typically limited to Commanding Officers or other senior personnel.

However, so many RSN submariners qualify on Swedish submarines that, upon qualification, many of them receive Swedish sharks from Sporrong as their first qualification badge, THEN receive the RSN badge. Singapore has worked out an arrangement with the Swedish embassy to arrange for the old-style Swedish badges with the serrated front to be made locally in Singapore, not by Sporrong, due to the larger number of submariners who qualify on Swedish submarines (SWE-4-SIN). These additional badges are worn on the upper right chest of the uniform.

SWE-4-SIN

A patch is authorized for wear on the RSN submarine coveralls. Rectangular, it has a gray-blue background with the submarine badge design in the center (SIN-1C.2).

SIN-1C.2

SOUTH AFRICA (SAF)

Though having a long seafaring history, South Africa did not join the list of countries with submarines until they completed the purchase of three *Daphne* class submarines from France in April of 1967. With the official launching of their first submarine on 18 March 1969, the South African Submarine Service was created, becoming the first country in Sub-Saharan Africa with this capability.

Eager to have a South African submarine distinctive, two designs were featured in the SA Naval News, a full year prior to the first submarine being commissioned. Both designs featured inward-facing heraldic dolphins, one supporting an interior shield and the other a wreath (SAF-SKETCH-1.1 & 1.2), however neither of these designs was selected. On 16 January 1970, the first examples of the newly approved South African submarine badge were issued by Commander J.A.C. Weidman. This badge was a simple gold (gilt) silhouette of a stylized *Daphne*-class port-facing submarine (SAF-1). It was worn on the right breast of all uniforms and was issued to all ranks following a qualification process taking approximately two years for most submariners.

SAF-SKETCH-1.1

SAF-SKETCH-1.2

There were three variants issued of this first South African badge. The first was a "casting" (SAF-1.1*), the second was a thick "heavy" pressing (SAF-1.2), and the third was a "light" pressing with a hollow back (SAF-1.3). All three variants look <u>identical</u> from the front, the differences only noticeable from the rear and sides. All have clutch-pin attachment methods.

SAF-1.2

SAF-1.3

Though the uniform regulations provided an allowance for a mess-dress size, no mess-dress or cloth versions of this first badge are known to exist. Because of the relatively brief period the SAF-1 was issued (1970 - October 1974), and the fewer than 200 recipients, it can be difficult to find a verified version of this badge. As the three original variants were made of brass, a private effort created a small number of the SAF-1 badges with gold plating to resist tarnishing (SAF-1.4*). The exact quantity of these issued is not known by the authors.

A warning to collectors – multiple unauthorized copies of the SAF-1 badge have been created, some with horizontal pin back and some with clutch back. These copies can make it challenging to identify the original versions. Additionally, in 2015 approximately 50 re-strikes of the SAF-1.3 "light pressing" were created from the original dies (SAF-1.5*). Because these can be so difficult to tell apart, provenance is key.

The reason the SAF-1 badge was issued for such a brief period was that it was unpopular with the SAF submariners. In late 1973, a movement began within the submarine community to drive design of a new badge. A prototype of the proposed new design was created by one of the best jewelers in South Africa, and was widely accepted by the submarine force. This badge is now held by the Senior Officer of the Submarine Service, and is passed down to subsequent holders of that position. On 25 October 1974, the second design of the South African submarine insignia was authorized and issued. This badge was a port-facing gilt *Daphne*-class submarine in front of a silver trident, surrounded by two pewter laurel branches (SAF-2).

Like the SAF-1, this badge was also worn on the right breast, and the same badge was worn by all ranks. This design is two-part construction with the submarine profile and the wreath being separate pieces, but both secured with the same two clutch pins. This badge also had multiple manufacturing variations which we will describe in detail, but will not show. The first is an all pressing with short clutch pins and a wreath that wears to a copper/brass color (SAF-2.1*). The second has the same pressed gilt submarine with an antiqued pressed wreath and trident with long clutch pins (SAF-2.2*). A third version has a pressed, plated submarine that wears to a silver color, over a wreath that wears to a dull grey color (SAF-2.3*). A fourth version has a brass-plated submarine and solid antiqued wreath and trident, with larger holes in the wreath, and short clutch pins (SAF-2.4*). The last version is exactly the same in that it has the same plated submarine, but has a solid cast anodised wreath and trident, one version has short clutch pins (SAF-2.5S*), and the other has long clutch pins (SAF-2.5L*). It is important to note that no genuine badges were made with four clutch pins on the back… two each on both the submarine and the wreath/trident. Badges with the four clutch pins are known fakes. Two examples are shown below (SAF-2-EX1 & SAF-2-EX2).

SAF-2-EX1

SAF-2-EX2

In the 1980s, an attempt was made to implement new cloth patches on denim material for working uniforms and jackets. While a few prototypes of these badges were created, they were never approved for use (SAF-2-UAC.1 & UAC.2).

SAF-2-UAC.1

SAF-2-UAC.2

As early as 1986 efforts were implemented to change the SAF-2 design for a new style badge somewhat similar to that shown in SAF-SKETCH-2. This effort finally gained some traction in the mid-1990s when a design similar to the 1986 sketch was created and submitted for approval.

SAF-SKETCH-2

This badge had two grades, gold for Commanding Officers (SAF-3.1) and silver for all other officers and crew (SAF-3-UA1). Each had two inward-facing bottle-nosed dolphins facing a compass rose with an upward facing vertical dagger in the center. They were constructed of two parts that screwed together, and were made in a full size and mess dress version (SAF-3.1m* and SAF-3-UA1m), both with maker's marks and clutch pin attachments. Full-sized cloth versions were also made in both gold (SAF-3C.1) and silver (SAF-3-UA1C). After all of the manufacturing and changes to the uniform regulations were completed, just prior to the badges being presented at a parade in 1997, the design was officially rejected. This was done in large part due to the badges not meeting the original intent of the submarine insignia, which was to provide a symbol of unity (no ranks or grades) and distinctiveness.

SAF-3.1

SAF-3.1r

SAF-3-UA1

SAF-3-UA1r

SAF-3-UA1m

SAF-3-UA1mr

SAF-3C.1

SAF-3-UA1C

So, since these two badges were never officially approved as a submarine distinctive, why are they being included in this book? The reasons are three-fold: First, rarely have badges gone through the entire design and approval process to the point of having uniform regulations created, only to be rejected, literally, just prior to release. Second, we show these to highlight the differences between them and the 2009 badges… and another reason listed below.

In 2009, following the launching of three newer Type *209* German-made submarines, the South African Navy decided (for real this time!) to replace the SAF-2 badges with a new style issued only in gold for all ranks to coincide with their new submarines. Adopting the 1997 badge design, the initial issue made use of the earlier production of the SAF-3s & SAF-3m badges from 1997… therefore, these previously rejected badges actually reached approved status. Regarding the rejected silver dolphins (SAF-3-UA1 & SAF-3-UA1m), they were given to veterans attending one of the submarine service reunions as a momento, so did not go to waste.

The supply of the original SAF-3.1 badges quickly dwindled as only 100 of each had been made, and new stock had to be created for full-size and mess dress badges, but by a different manufacturer. These new badges were similar to the design of the SAF-3s and also similar in their two-part screwed construction, but with some noticeable differences. First, there is no maker's mark on these badges, and where the SAF-3.1 had the noses of the dolphins slightly below the bottom circumference of the compass rose, the new badges had the dolphin's noses resting on the circle, with a slightly different dagger (SAF-3.2 and SAF-3.2m).

SAF-3.2

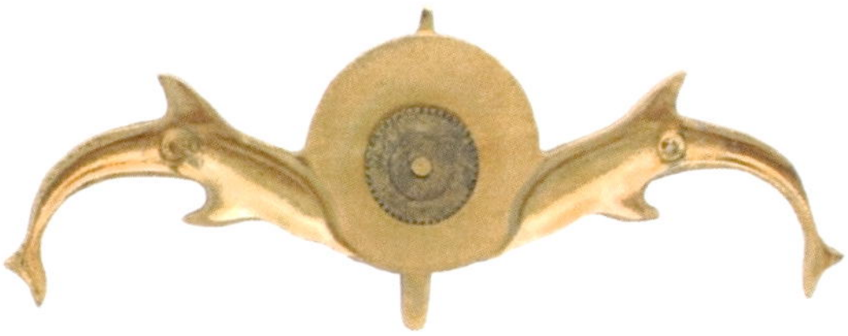

SAF-3.2r

SAF-3.2m

SAF-3.2r

There is a variation of this badge that is single-piece construction with a smooth back (SAF-3.3), also with no maker's mark. All of the later SAF-3.3 variations have clutch back attachment. Different from the SAF-1s and SAF-2s, the SAF-3 badges are to be worn on the left breast above the ribbon bar, not the right.

SAF-3.3

SAF-3.3r

Neoprene versions of this design were created for wear on work uniforms, with a gold on black design (SAF-3C.2), and a blue on white (SAF-3C.3). A black on green camo (SAF-3C.4) was issued for medics that differs from the other neoprene badges by having clutch back attachments. The SAF-3C.4 is quite rare with fewer than 20 badges created.

SAF-3C.2

SAF-3C.3

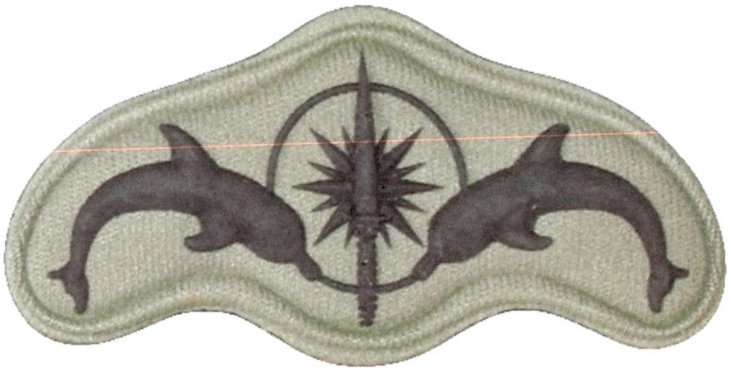

SAF-3C.4

There are multiple unofficial examples not to be worn on uniforms of all three designs created for various uses, such as tie tacks and lapel pins that will not be addressed here, and are available in other reference materials.

SOUTH KOREA / REPUBLIC OF KOREA (KOR)

Though operating midget submarines since the 1980s, the Republic of Korea (ROK), also known as South Korea, did not have a full-sized submarine until purchasing one from Germany in the early 1990s. After that initial purchase, all ROK submarines have been constructed in South Korean shipyards.

A Republic of Korea Navy submarine qualification badge was first authorized in 1992, and, to our knowledge, did not previously exist for the midget submarines. Following the approval process, on 6 February 1993, the newly designed badges were presented to the officers and crew of the first full-sized ROK submarine, the *Chang Bogo*, while it was being commissioned in Germany.

This badge is a bow-facing modern submarine atop a naval anchor with pierced arms and a Korean Yin/Yang symbol on the bow. The submarine is flanked by two scaled bottle-nosed dolphins, their noses connected to the sail, with waves from the submarine's wake connecting it to the dolphins. It is gold for officers (KOR-1) and silver for enlisted personnel (KOR-2) with clutch back fasteners. Cloth versions of gold (KOR-1C) and silver (KOR-2C) thread on black wool were also issued. The South Korean dolphins are awarded after 3 to 6 months of classroom training, followed by written and practical exams to be completed within 7 months aboard a submarine. Upon completion the captain orders a ceremony to award the dolphins. The badge is worn on the left chest of all uniforms, above the ribbon bar if present.

KOR-1

KOR-2

KOR-1C

KOR-2C

By 1996, the South Korean submariners had become dissatisfied with their dolphins, so Admiral An Pyong-koo, captain of the *Chang Bogo* designed a new badge that was much more sleek and widely accepted by the submarine community.

The new version also featured a modern submarine in the middle, but without the anchor or the Yin/Yang symbol. The submarine hull is supported by the tails of two smooth bottle-nosed dolphins swimming up and away from the submarine. This badge was also in gold for officers (KOR-3) and silver for enlisted personnel (KOR-4). Early versions of these badges were known to have screw back fasteners, and newer versions have clutch-back attachment points. There are miniature versions of these badges that are not officially issued and are worn only with civilian clothing. Cloth versions using gold (KOR-3C) and silver (KOR-4C.1) thread were also made and used through 2020.

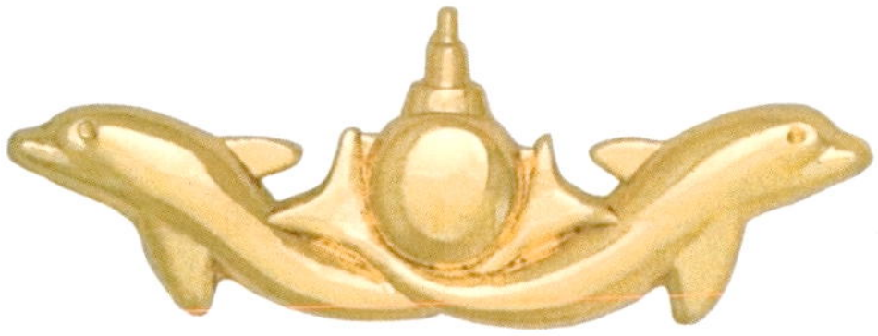

KOR-3

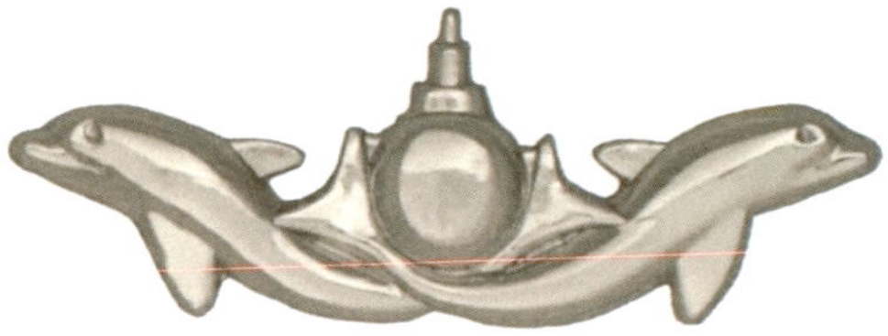

KOR-4

KOR-3C

KOR-4C.1

In 2021 the cloth badges were changed to reflect new unforms. Both officer and enlisted personnel use patches with silver dolphins having a different background dependent upon the uniform used. The ship's unform cloth badge is silver on dark blue (KOR-4C.2) while the combat uniform badge is silver on olive drab (KOR-4C.3).

KOR-4C.2

KOR-4C.3

SPAIN (SPN)

Spanish inventors had some of the earliest successful efforts with submarines in the 19th century, and though the success of some such as Isaac Peral resulted in serious consideration of including submarines in their fleet, it took some time to get submarines officially added to the Spanish fleet. First ordered in 1908, then again in 1912, neither effort succeeded. Finally, a royal decree issued 17 February 1915 authorized the purchase of four submarines, the first one a *Holland*-class submarine purchased from Electric Boat in the United States, arriving on 25 April 1917, and christened as the *Isaac Peral.*

Upon receiving their first military submarines, the Spanish Naval Command, spearheaded by the commander of the Spanish submarine fleet, Mateo Garcia de los Reyes, supported developing a unique distinctive for the officers and sailors serving aboard submarines. When presented with the idea, King Alphonse XIII agreed and by royal decree on 5 September 1919, the Spanish submarine insignia was approved for issue.

This first badge was port-facing detailed profile of a Spanish *A*-class gold submarine with a colored oval in the center and a detached royal crown of Alphonzo XIII above the submarine. This badge was made as a gold metal badge and also as a cloth patch with gold filament. The oval in the center of the early metal badges was cloth – canvas by some accounts. This badge was worn by all submarine personnel, on the right breast of officers and senior enlisted personnel, and on the right sleeve of other enlisted personnel.

There were two versions of the cloth patch, one made with a red center for all submarine-qualified personnel (SPN-1C), and the other with a blue center for submarine school trainees not yet assigned a submarine billet (SPN-2C*). The metal badge had a removable disk that held the interchangeable colored cloth in place, easily allowing the transition to a qualified submariner badge (SPN-1.1 & 1.2) from a trainee badge (SPN-2*). Valid blue versions are incredibly rare and seldom seen because most trainees went on to full qualification.

SPN-1C

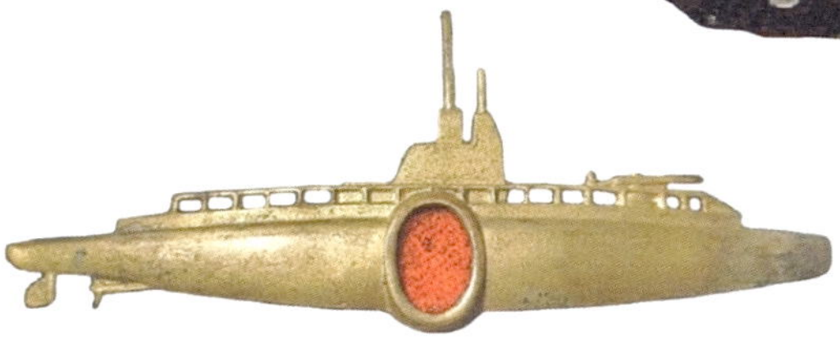

SPN-1.1

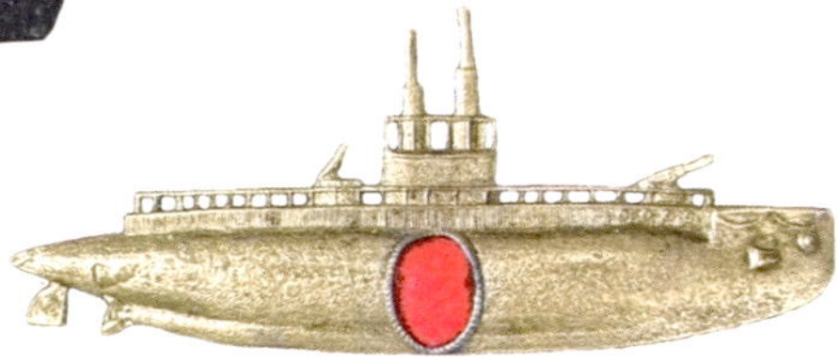

SPN-1.2

In April 1931, following the fall of the Spanish Monarchy, the already detached crowns were officially removed from these badges, technically making them a different badge. While the submarine profile largely remained the same, new submarine profiles began to be introduced. Soon after, the 2nd Republic leadership decided to implement the use of a mural crown on badges, including submarine badges (SPN-3 (SKETCH) & SPN-3C).

SPN-3 (SKETCH)

SPN-3C

The Spanish Civil War that occurred from 1936-1939 threw Spain and the Spanish military into chaos. The existing Spanish submarines were retained by the sitting government Republicans, while submarines from Germany and Italy fought for the Nationalist opposition led by General Francisco Franco. Two Italian submarines were sold to the Nationalists and crewed with Italian officers and Spanish crews at first, transitioning to a fully Spanish compliment by the end of the war. Four Italian submarines were loaned to the Nationalists, crewed by mostly Italians, with a few Spaniards. These crews sailed as part of the "Spanish Legion." As with so much information lost to the fog of war, many details regarding the submarine insignia associated with the combatants is less than clear.

SPN-REP-1

While it is generally believed the Republican submarine crews maintained use of the SPN-3 badges, there is evidence that indicates some cloth badges were created for Republican submarine crews with a Soviet flavor due to the Spanish Government seeking assistance from Soviet Union at the time. Reproductions of these badges are shown (SPN-REP-1, 2, 3, & 4) since the authors do not have permission to show photos of the originals.

SPN-REP-2

SPN-REP-3

SPN-REP-4

There are several badges known to be associated with the Nationalist Spanish/Italian hybrid crews. The first may be one of the rarest badges in this book. This gilt badge is an unusual starboard-facing stylized submarine, with a cable cutter on the bow, a low-profile sail, fore and aft deck guns and an Italian Bourbon crown embossed in the middle of the submarine. The submarine profile is superimposed over a Spanish Legion crest of a trio of crossed weapons – a ceremonial halberd (pike), a musket, and an arbalest (medieval crossbow) (SPN-NAT-1). Sources indicate examples hallmarked by the jeweler Fermin exist. A cloth version of this badge has the badge outline and crown in red thread on a dark blue background (SPN-NAT-1C).

SPN-NAT-1

SPN-NAT-1C

Another badge from this era features a similar style gilt submarine profile, only facing to port, surmounted by the Italian Bourbon crown, but lacking the Legionnaire's crest. Two versions of this badge are known. The first is a plain version, just as described above, with no known maker's mark or jeweler's hallmark (SPN-NAT-2). The second has a very similar profile and crown, but in addition, has circular embossment in the center with the few known examples having either a red stone or red paint inside. This badge has a Fermin maker's mark on the back, and also a "916/000" assay mark with a hexagonal jeweler's hallmark used between 1934-1988 that indicates it is of Spanish origin (SPN-NAT-3). On these badges, the gilt has been poorly preserved and much of the silver base metal shows. The distinction between these badges and their purpose are unknown, other than they were likely issued to crews on the Italian submarines fighting for the Nationalists. They are included here in the Spanish section due to being created for submariners fighting in the Spanish Civil War.

SPN-NAT-2

SPN-NAT-3

SPN-NAT-3r

Following the end of the Spanish Civil War and with the rise of the Franco regime, a new submarine badge was issued around 1940 that had a newer style submarine that was modeled after the two Italian submarines that were sold to the Nationalist forces during the Civil War. Less stylized than the Italian-made badges, the port-facing submarine profile is more angular with more details etched into the sail and hull. This badge also features the open-topped royal crown adopted by the Franco regime with a colored "jewel" at the center. The regulations for it were issued in 1941.

The jewel color in these badges was associated with the qualification level attained by the individual. Red jewels were for those qualified and assigned to submarines (SPN-5). Green stones replaced the previous blue center for those passing the submarine aptitude test and in training (SPN-6). Black stones were for personnel serving at least two years in submarine support positions (SPN-7). There are many jeweler and theater-made versions of the SPN-5. One such example is shown (SPN-5J) that has a much more detailed crown and appears to be from an earlier time period – possibly the 1940s-1950s. The "official" badge design did not have a cable cutter on the bow, however, multiple examples exist with and without the cable cutter. There are mess dress versions of this badge with both port (common) (SPN-5m-P) and starboard-facing (less common) (SPN-5m-S) profiles. Cloth versions are also known of the full-size badges (SPN-5C, SPN-6C* & SPN-7C*).

SPN-5

SPN-6

SPN-5J

SPN-7

SPN-5m-P

SPN-5m-S

SPN-5C

The Spanish badges were not modified again until after the end of the Franco era. Following the death of Franco in 1975, the crowns were changed quite quickly from the Franco open-topped royal crown to the crown of newly anointed King Juan Carlos, while the submarine profile remained the same as the previous version. This change was not made official until 7 November 1978, when government ministerial decree #990 was issued marking the transition. The color of the stones and their meanings remained the same (SPN-8, SPN-9, SPN-10). There are multiple variations of these badges. Versions are known with a solid gold crown, a solid crown with red enamel, and pierced crown (SPN-8.1, SPN-8.2, & SPN-8.3), and also some early examples with the cable cutter on the bow (SPN-8.4). Cloth and bullion versions of these badges were also implemented (SPN-8C, SPN-9c*, SPN-10c*).

SPN-8.1

SPN-9

SPN-8.2

SPN-10

SPN-8c

SPN-8.3

SPN-8.4

Like the Franco badges before them, mess dress versions of all three levels of these Juan Carlos crown qualification badges were approved for use (SPN-8m.1, SPN-9m, SPN-10m), including different crown configurations (SPN-8m.2), and a starboard-facing version (SPN-8m-S).

SPN-8m.1

SPN-9m

SPN-8m-S

SPN-10m

SPN-8m.2

The post-Franco era badges remained in effect until 1986 when the badges were changed to have a submarine profile similar to the SPN-1. The Juan Carlos crown was retained and filled with red enamel. The submarine support position badge was officially eliminated, and only the badges for the qualified submariners and trainees remained (SPN-12.1 and SPN-13.1) with a few variations (SPN-12.2 & SPN-13.2). Cloth and bullion versions (SPN-12C & SPN-13C) and jeweler's versions exist (SPN-12J & SPN-13J). Though the support badges were eliminated, examples of personnel wearing them can still be found. These were the last changes to the Spanish submarine badges to date. Miniatures were not created since full-size badges are approved for use on mess dress uniforms.

SPN-12.1

SPN-13.1

SPN-12C

SPN-13C

SPN-12.2

SPN-13.2

SPN-12J

SPN-13J

SWEDEN (SWE)

The Swedish submarine service began in 1904 with their purchase of His Majesty's Swedish Ship (HSwMS) *Hajen* (Shark), a Swedish-built submarine based on the USS *Holland* design. This began a long and distinguished Swedish history of underwater ship building and operating.

As early as 1908, badges were proposed for enlisted personnel. Designs were submitted by the Navy Chief of Staff on 3 March 1908, and forwarded to the Department of Naval Affairs. Suggestions included two different types of a vertical trident, one three-dimensional and the other flat, with various symbols surmounted in front; examples included a ships wheel for non-commissioned officers (NCOs), a propeller for engine room personnel, and crossed semaphore flags for boatswains. However, these designs were never approved, so were never created.

From 1910 through 1943 designation patches were approved for junior enlisted personnel serving aboard submarines. For personnel working in the engine room, a yellow propeller on a black background with a "U" at the top (for *Ubåt*) for winter blues (SWE-1.1), and blue on white for summer whites (SWE-1.2). For the rest of the junior enlisted personnel, patches were issued with crossed flags featuring the "U" above them. These were issued with yellow flags on a black background (SWE-2.1) and blue on white (SWE-2.2) for the enlisted personnel, and with blue flags on black for conscripts (SWE-2.3). The yellow flag version was also issued with an "A" below the crossed flags for "Artillery Observer" (SWE-2.4).

SWE-1.1

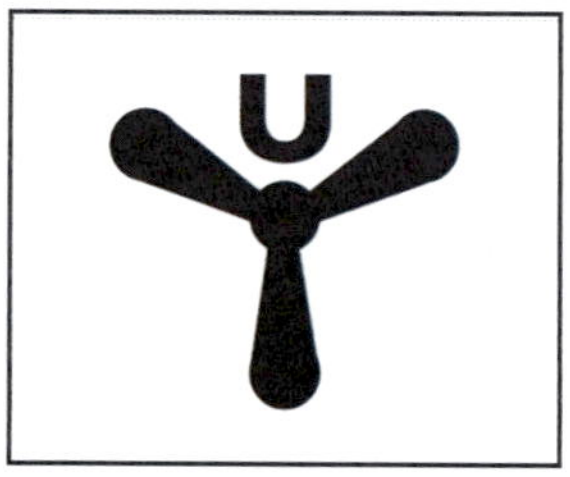

SWE-1.2

SWE-2.1

SWE-2.2

SWE-2.3

SWE-2.4

In 1940 personnel in Swedish Submarine Flotilla 6 had a badge designed. A bronze port-facing submarine transiting on the surface above an anchor in the center of a flotation device with the words "FOR SVERIGE UB. VI" flanked by diagonal years, "1939" on the left and "1940" on the right (SWE-FLOT-1*). This was not used as a qualification badge and was flotilla designation only.

On 16 March 1944, a uniform committee submitted a proposal for new crew uniforms and insignia to the Chief of the Navy. This included a series of prototype badges for all naval services, including submariners. This badge was a port-facing submarine profile above two stylized waves captured in a circle, mounted on a narrow bar with angled ends that have holes for sewing to the uniform (SWE-PTYP*). The only known photo of this badge is from a Swedish military magazine. The entire series of badges was relegated to the scrap heap by 1948 as it was deemed to be "too German" too soon after WWII and never approved for wear. A reproduction of this badge has been manufactured, with the submarine made of silver metal, on a dark blue enameled bar (SWE-PTYP-R). The colors, measurements, and materials of this badge were verified based upon comparison with a known badge for a different designation. The number of badges originally produced as prototypes is unknown.

SWE-PTYP-R

In 1956 an unauthorized submarine badge was created by Swedish submariners and produced by the Swedish badge manufacturer, Sporrong. Called "Dolphin" in English by the Swedish submariners, the badge shows a gilt-colored sail section of a port-facing submarine complete with limber holes, periscope, and snorkel, with two stylized waves crashing toward it, and a large trident fork superimposed on the front (SWE-UA.1). All original versions of this badge have Sporrong's maker's mark on the back. There are known examples of badges with the exact same dimensions as the originals that appear to be period pieces, but without the Sporrong maker's mark (SWE-UA.2). Sources say Sporrong has verified also making these badges. A reproduction of this badge was made in the mid-2000s but is slightly smaller than the original. Both the SWE-UA.1 & UA.2 are horizontal pin backs, but the reproduction has a roller catch while the originals have a simple hook catch.

SWE-UA.1

SWE-UA.1r

SWE-UA.2

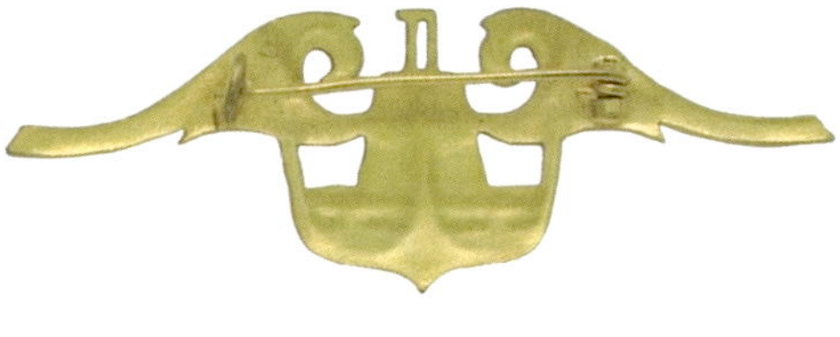

SWE-UA.2r

In 1976 the Chief of the Navy requested the State Heralds (official seal designers) create a submarine qualification badge. Surveying the submarine crews, a design was submitted with sharks (SWE-ORIG-SKETCH) as a nod to the first Swedish submarine, the *Hajen*. This design was originally rejected because it was not heraldic compliant, and the Heralds suggested two alternate designs (SWE-HER1-SKETCH & SWE-HER2-SKETCH). After negotiations, the final design submitted by the State Herald, Jan von Konow, was approved (SWE-APP-SKETCH). Based upon this approved design, an officially authorized insignia was finally introduced 6 September 1978. This was a cloth insignia with gold thread on a dark blue background depicting a central shield with Neptune's trident supported by the tails of two outward-facing sharks and topped with the Swedish royal crown (SWE-3). Swedish submariners earn their "sharks" following completion of their training and spending at least six months aboard an operational submarine and qualifying as watch standers. The badge is worn by all ranks on the right chest above the pocket of all uniforms.

SWE-ORIG-SKETCH

SWE-HER1-SKETCH

SWE-HER2-SKETCH

SWE-APP-SKETCH

SWE-3

A metal version of this badge was introduced in 1986. It had the same design as the cloth version, but it was made of gilt metal and the sharks had a scored/serrated finish on the front. These were made by Sporrong, and the first 100 were numbered (SWE-4.1) and given to only officers, and then an additional 399 numbered badges were made and presented to the entire submarine force. Sporrong continued to make the badges even when numbering was no longer required, and these are easily identified by the "Sporrong" maker's mark on the back and no number (SWE-4.2). At some point after the numbering stopped, Sporrong added a scaled appearance to the backs of their badges, further uniquely identifying them. There are also versions of this badge that are not made by Sporrong, and it is thought that these are not approved for wear by Swedish submariners. There are minor design differences for these with less pronounced gills, thicker and less distinct forks on the trident, less detail on the crown, and a smooth back, not scaled. These may be the version made in Singapore, but that has not been confirmed (SWE-4-SIN? in the Singapore chapter). It is important to note that, while widely used and tolerated for wear by the naval authorities, the SWE-4 badge was <u>never</u> officially authorized as it did not meet Swedish heraldry qualification standards.

SWE-4.1

SWE-4.1r

SWE-4.2r

The SWE-4 was in use until late 1998 when a new, slightly modified badge meeting heraldry qualification standards was approved for use. Instead of the scored surface, the new sharks are smooth, and while the shield in the center still features Neptune's trident, it now emerges from the sea at the bottom of the shield (SWE-5). This has the effect of making the shield taller than the previous version. These are also made by Sporrong, and every official badge has the scaled back and the Sporrong maker's mark.

SWE-5

SWE-5r

The 1998 change also resulted in another new badge being approved for use. A copper-colored badge of exactly the same design was created for conscripts, or personnel in training status aboard submarines (SWE-6), with the gold badge being reserved for those completely submarine qualified. A few years prior to this badge being approved, Sporrong created a very small number of copper-colored badges (around 10) using the old SWE-4 mold to assess how a proposed conscript badge would look (SWE-6-PTYP). The SWE-3, SWE-5 & SWE-6 remain the standard issue badges at the time of publication.

SWE-6

SWE-6r

SWE-6-PTYP

There is a commanding officer's badge that was designed and scheduled to be issued in conjunction with the other 1998 badges. This badge is gold in color and the same basic design as the standard badges, with the addition of a laurel wreath around the center, and an upward-facing sword piercing the shield and crown (SWE-CO-UA). Interviews with Swedish Naval personnel have determined that though a small number of these badge were issued by Sporrong, it was never officially authorized due to the submarine flotilla commander vetoing the badge because, "all submariners face the same perils, so should have the same badges." As with other Swedish badges, counterfeit examples exist.

SWE-CO-UA

SWE-CO-UAr

There are several special Swedish badges that are worth mentioning here. At some point prior to 1998 a new silver badge was created. This badge was exactly like the SWE-2, except it was silver (SWE-7.1). As the story goes, a surface warfare officer (AKA, a "Skimmer") was selected to lead the submarine squadron. Since he was not qualified to wear sharks, his staff took the initiative to suggest the creation of silver sharks for presentations to VIPs, as the squadron commander sees fit. After approval, it was later expanded to include officers in charge of submarine support functions. This custom apparently continued into the 2000s as smooth silver sharks are known (SWE-7.2). Again, official versions of this badge are made by Sporrong, and badges without that maker's mark are fakes.

SWE-7.1

SWE-7.1r

SWE-7.2

SWE-7.2r

In 1996, the *HSwMS Gotland* crew and the *Ubåtsräddningsfarkost* (URF), or Submarine Rescue Vehicle participating with the *Gotland* created special issue "Polar Voyage" badges as a remembrance of a training exercise that took place within the Arctic Circle. The badge is the same design as the post-1998 sharks, but instead of Neptune's trident emerging from the ocean, the shield is dark blue enamel with a light blue globe and white polar ice cap pierced by a gold trident, and the crown has red enamel fill. These are ceremonial only and cannot be worn as a qualification badge (SWE-PV). A mini version (SWE-PVm) and a tie bar (SWE-PV-TB) were also made and these were issued as a numbered set. Only about 60 of these sets were issued.

SWE-PV

SWE-PVr

SWE-PVm

SWE-PVmr

A unique aspect of the Swedish submarine force is their "Quad Club". A submariner originally joined this elite group by meeting the following criteria:

- Have sailed on 4 submarine types
- Be over 40 years of age
- Have fired over 400 torpedoes
- Have 4,000 days of submarine service

When qualifying as a Quad Club member, submariners get special "colorized" sharks that have the red highlights of the royal crown filled in, the trident in the shield is black and the water of the ocean is a royal blue. The Quad Club officially started in 1994, so there are SWE-4 examples with the highlights (SWE-4QC) as well as SWE-5 (SWE-5QC). There are also Quad Club versions of the unauthorized commanding officer badge which, of course, have no official standing (SWE-CO-UA-QC). Due to the incredibly stringent original criteria for club entry, it is almost impossible for the current Swedish submariners to meet the standards. So, the rules have been changed and now only require the entrant to be greater than 40 years old and to have served 4,000 days aboard submarines.

SWE-4QC

SWE-4QCr

SWE-5QC

SWE-5QCr

SWE-CO-UA-QC

SWE-CO-UA-QCr

Though not an official badge, one of the Swedish submarine veteran's associations, Nordkaparen, has issued a club badge. This badge is a miniature size and utilizes the sharks of the qualification badge but replaces the trident and ocean in the center with the coat of arms of the city of Gothenburg, a heraldic lion on a field of blue and white diagonal stripes (SWE-NORD). Other Swedish submarine veteran organizations exist, some with badges, most without.

SWE-NORD

SWE-NORDr

Similar to other countries, Sweden has unofficial boat badges that are used by crew members of the individual submarines. These pins were first used in the early 1990s, originating from Swedish Army regiments that represented various regions in Sweden that had these symbols as their official crest. The various Swedish submarines named after cities in these regions adopted these crests as their unofficial symbol over the years. Some examples are shown here.

HMS
Halland

HMS
Södermanland

HMS
Gotland

HMS
Uppland

HMS
Östergötland

HMS
Hälsingland

Since the issue of the first official Swedish badge, other patches have been utilized, typically on working or underway uniforms. These were initially issued with gold sharks around early 2005 (SWE-5C.1 & 5C.2). Since then, a green badge has been issued for use on other uniforms (SWE-5C.3).

SWE-5C.3

SWE-5C.1

SWE-5C.2

Patches and badges are also issued showing the trident surmounted by the crown that is the official insignia of the Swedish submarine service. Anyone associated with the service is entitled to wear these, not just personnel who are submarine qualified (SWE-TRI-1 & SWE-TRI-1C.1 – 1C.5).

SWE-TRI-1C.1

SWE-TRI-1C.2

SWE-TRI-1C.3

SWE-TRI-1C.4

SWE-TRI-1C.5

SWE-TRI-1

More recently, patches for the submarine service have been issued with the same basic design, but no crown (SWE-TRI-1C.6).

SWE-TRI-1C.6

While only for use on civilian clothes or as tie bars and not approved for official use, miniature badges exist for SWE-4NN, SWE-5, SWE-6, SWE-CO-UA, SWE-7.

TAIWAN/REPUBLIC OF CHINA (TAI)

The Republic of China (ROC), also known as Taiwan, began their submarine service in late 1969 with the purchase of two Italian *SX404*-class midget submarines. These were decommissioned in 1973 when Taiwan purchased a *Tench*-class and a *Balao*-class submarine from the U.S. Navy. The *Tench*-class submarine, the ROCS *Hai Shih*, currently has the distinction of being the world's longest serving submarine and is expected to continue to serve until at least 2026.

The ROC Navy first issued qualification insignias on 8 October 1969 when the Italian midget submarines were commissioned as the ROCS *S-1* (*Hailong*) and *S-2* (*Haijiao*). The badge has two inward facing scaled heraldic dolphins and a forward-facing round-hulled submarine in the center, with the round blue sky and silver 12-pointed sun symbol of the ROC Naval Jack featured on the bow of the submarine. This badge is issued to officers in gold (TAI-1.1) and to enlisted personnel in silver (TAI-2.1). It is worn on the left chest above the ribbon bar. This same basic design is still in use, though there have been some minor changes to how the national symbol in the middle is fabricated, and materials used for the badge.

TAI-1.1

TAI-2.1

The early badges were fabricated from aluminum and had double screw-back attachment. They were solid back and had six Chinese characters on the back signifying "Republic of China Navy Submarine Chest Insignia", and for the badges initially issued to individuals (their first badge), both officer and enlisted badges were numbered, some with a letter "S" in front and some without the "S" (TAI-1.1r & TAI-2.1r). Replacement badges were available with the writing, but without the number. There were VIP badges issued only in gold, also numbered and identical to the officer's badge, but with 12 Chinese characters on the back stating "Republic of China Navy Honorary Submariner" (TAI-1.1VIPr). For these early versions, the central sun symbol was a metal-based photoetched appliqué.

TAI-1.1r

TAI-2.1r

TAI-1.1VIPr

It was noted that the central appliqué on the first versions could experience significant fading over time, so modifications were made to subsequent versions. The second version had a cloisonne central sun symbol that avoided the fading issue (TAI-1.2 and TAI-2.2). According to sources this badge may have only been in use for a few years. Versions of this badge are known that are solid aluminum and others of a more robust metal that are hollow back. Some of the solid back badges have different wording on the back with eight Chinese characters saying, "Command Department of the General Staff", and some are also numbered (TAI-1.2r).

TAI-1.2

TAI-2.2

TAI-1.2r

Around 1990, the ROC Navy issued a change to all badges carrying the national sun symbol. The Kuomintang political party had adopted a rather large sun very similar to those on the ROC Navy insignia, so the military changed the symbol to reflect a much smaller star, more similar to the national symbol. So, the third and most recent version has a much smaller central sun symbol in relief and the badges are all made of the sturdier metal (TAI-1.3 and TAI-2.3). This badge is known in both hollow and solid back variants. This is possibly due to submariners removing the previous style central symbol from the badge and replacing it with the new center.

TAI-1.3

TAI-2.3

Miniature badges of this latest version are authorized for use with mess dress uniforms (TAI-1.3m and TAI-2.3m).

TAI-1.3m

TAI-2.3m

Cloth versions of both the officer and enlisted badges exist that are authorized for use on coveralls (TAI-1.3C and TAI-2.3C). The year of authorization for these is not known by the authors.

TAI-1.3C

TAI-2.3C

THAILAND (THA)

Thailand (then Siam) created their Navy in 1906. Though the Royal Siamese Navy (RSN) had a desire for submarines almost immediately, there was not a budget to support this effort until 1936. At that point, the RSN procured four *Matchanu* class submarines from Mitsubishi in Japan. These were delivered to the RSN on 4 September 1937 (still celebrated as "Submarine Day" in Thailand), commissioned 19 July 1938, and were in service until 1951 when they were decommissioned and scrapped, save for the conning tower of the *Matchanu* that is now located at the Thai Naval Museum.

Though details are lacking regarding the possibility of a submarine badge, two possible scenarios have been identified. The first is that the sailors aboard these four submarines received submarine qualification training in Japan and were supposedly each presented with a hand-crafted submarine qualification badge. The other, and possibly more likely story, is that several officers receiving OJT on the submarines in Japan designed a badge to be worn among themselves. This badge shows a starboard profile of a submarine with cable cutter and a bow anchor attached with a chain, surrounded by a laurel wreath open at the top with seven lotus blossoms (THA-1-PHOTO). The only details known about this badge are from a poor black and white photocopy of a photograph that was discovered, so no information is available on the badge details such as the materials, size, colors, and attachment method. To date, no known original versions of this badge have come to the attention of the collecting community, making it possibly the rarest of all submarine insignia. A high-quality reproduction has been fabricated based on the photograph (THA-1REP).

THA-1-PHOTO

THA-1REP

In the early 2000s there was increased interest in reestablishing a submarine service within the Royal Thai Navy (RTN). By 2011 examples of a proposed badge began to be circulated, and in 2017 the Thai government reached a deal to purchase three diesel-powered submarines from China. At the time of publishing, this acquisition is still not complete, but official sources have confirmed that the proposed badge (THA-PROP-1) that was around for several years was not accepted as the new RTN submarine badge and therefore has no official capacity. Copies of this badge have been created in silver and gold, though, according to sources, only the gold version was proposed.

THA-PROP-1

The badge that has been accepted as the new official RTN submarine qualification insignia is designed with a nod to the original badge and additional symbolism.

- The port-facing HTMS *Matchanu* class submarine profile represents the history and identity of the Royal Thai Navy's submarine force
- The cradle of boughs from the chaiyapruek flower symbolizes prosperity
- The length of the badge is 81mm, for the year 1938 when the Royal Thai Navy submarine force was created (Year 2481 in the Thai B.E. calendar)
- The badge is copper colored representing concealment – a vital capability for submarines

Upon qualifying as a Thai submariner, the candidate receives this badge mounted on a black velvet backing (THA-2). Badges without the backing are presented to individuals being identified as honorary submariners. This badge has a clutch back pin attachment method and is worn on the left chest of the uniform above the ribbon bar if worn. To date, this badge has only been earned by a few officers, so it is unknown if the same badge will be earned by enlisted personnel. The submarines are under construction and, due to recurring delays, the first is currently expected to be delivered in 2027.

THA-2

TURKEY AND THE OTTOMAN EMPIRE (TKY)

With the purchase of two *Nordenfeldt*-class submarines purchased from Great Britain by the Ottoman Empire in 1886 to defend themselves from a perceived Greek threat, the Turkish Navy became the earliest Navy to commission a submarine, with one of those submarines becoming the first to ever fire a torpedo under water. Though possessing a submarine captured during WWI, it was not until after the fall of the Ottoman Empire in 1922 and the creation of the Republic of Turkey in 1923 that they would once again venture into new submarine technology – and also submarine insignia.

On 24 September 1926, after sealing the deal to purchase two Dutch submarines, the Turkish Navy announced that a submarine insignia would be created for the sailors that would crew them. Soon after, an officer's badge was issued. This badge has a gilt oval wreath of oak leaves on the right, olive branches on the left, held together by a bow at the bottom and connecting with the Islamic crescent moon and star of the Turkish flag at the top. Across the middle of the badge lies a port-facing black and silver submarine profile (TKY-1). The submarine is attached to the wreath by a few pieces of bent wire, and the badge is secured to the uniform via a vertical pin. The back of the wreath is concave. These were worn on the left chest below the uniform pocket.

TKY-1

TKY-1r

Around the same time, NCOs assigned to these submarines were able to wear a ratings patch on their left shoulder featuring a submarine above their rank chevrons in gold thread on a dark blue background (TKY-2-SKETCH). Also, other enlisted personnel completing the submarine course were allowed to wear a submarine profile on a dark blue background on their left chest (TKY-3-SKETCH).

TKY-2-SKETCH

TKY-3-SKETCH

In 1928, when the new submarines were commissioned, a new metal badge was approved for senior enlisted personnel. This badge was made of brass and was a solid oval disk, once again with the oak leaves on the right and olive branches on the left, the bow at the bottom and the Islamic crescent moon and star at the top, but in the center is a starboard-facing submarine on the surface of a calm sea in front of a radiant sun with the letters, "T.C." above it for "*Türkiye Cumhuriyeti*" or "Turkish Republic" (TKY-4).

TKY-4

TKY-4r

The metal petty officer badge was in use until 1935 when all enlisted submarine insignia reverted to a cloth shoulder patch, but this time it was a combination of rank, rating, and service. This patch was gold thread on a black background like the previous version, but it was also available with gold thread on white material for summer whites and red thread on tan fabric (TKY-5.1, TKY-5.2, & TKY-5.3*). For this badge the submarine was under the chevrons indicating rank, and the rating symbol was above the chevrons. It is possible these were issued at different times, but this could not be verified by the authors.

TKY-5.1

TKY-5.2

During the late 1930s a German jeweler made some Turkish officer's badges. These differed from the Turkish-made badges in a few important ways. First, the wreath of the German-made badge is flat, not concave, is wider and taller than the Turkish version, and the crescent moon and star at the top are faceted on a solid disc. The submarine profile for the German-made badge has a two-step sail vs. a three-step sail on the Turkish badge. And most importantly, there is a maker's mark on the reverse of the German-made badge (TKY-1-GER). Multiple fakes of this badge exist.

TKY-1-GER

In 1956, the enlisted insignia reverted to a metal badge. This badge had a submarine profile similar to the one on the officer's insignia, but facing starboard, not port. This profile was mounted horizontally on a serrated oval wreath with two loops of a bow on top and the Islamic crescent moon and star affixed in the middle of the bow (TKY-6). This badge was secured to the uniform with a horizontal pin. It is believed to have been worn on the left chest of all uniforms. There is a verified version of this badge with a starboard-facing submarine profile, but it is not known if this is a unique example, or a possible variation.

TKY-6

TKY-6r

Only in use for three years, the TKY-6 was replaced with a new style badge in 1959. This new badge had a port-facing submarine profile, also similar to the officer's badge, but it was now mounted on a gilt oval of braided rope surmounted by three layers of ribbon, topped by the crescent moon and star (TKY-7). This was secured to the left chest with a horizontal pin.

TKY -7

TKY-7r

In 1964, another change was made to the enlisted submarine badge. The badge kept the same basic format, and the submarine profile remained the same, but the gilt oval braided wreath was slightly simplified, and the three bows on top were expanded and became more ornate, while the crescent moon and star remained the same (TKY-8). This badge was secured with a vertical pin and was in use until 1977 when all personnel were approved for wearing the original officer's badge (TKY-1). It should be noted that many reproductions/fakes exist of the TKY-4, 6, 7 and 8 badges including new versions of the TKY-4 and TKY-6 that are very well done and are being sold as originals.

TKY-8

TKY-8m

Multiple minor variations of the TKY-1 were issued over the years, and all were approved for use. In the late 1980s a mess dress badge was approved. Though not having the same level of detail, it was similar in every way, except for size, to the TKY-1. This badge was available in vertical pin and clutch back attachment (TKY-1m.1). In 1997 the current version of this badge was issued. Slightly larger and more well detailed, it is available in pin back only (TKY-1m.2). There are smaller lapel pins available, but these have no official standing. All three sizes are shown for perspective.

TKY-1.1

TKY-1m.2

TKY-1m.1

Fairly recently, patches featuring the Turkish qualification insignia have been approved for use on working uniforms. Various versions have been identified, including gold on royal blue (TKY-1C.1 & 1C.2), olive on camo (TKY-1C.3), black on denim (TKY-1C.4), and dark brown on khaki (TKY-1C.5). It is possible that some patches were in use prior to being officially approved.

TKY-1C.1

TKY-1C.2

TKY-1C.3

TKY-1C.4

TKY-1C.5

UNITED KINGDOM (UK)

Having a submarine history dating back to the successful operation of submarines built by Dutch inventor Cornelis Drebbel in the Thames River in the 1620s puts the United Kingdom at the forefront of viable submarine use, however it took several hundred years before they got serious about establishing a submarine force. Though the Royal Navy was at first very resistant to the idea, it was one of the initial adopters of modern submarines with their purchase of Holland-class submarines in 1900.

Like many countries, the UK had a well-established submarine force long before ever having an authorized submarine insignia. British submariners were first able to wear cap tallies during WWI. These were labeled simply with "H.M. SUBMARINE" and the submarine number, such as "E11". This was the standard until WWII, when the cap tallies were changed to read just "H.M.S." After the war they were changed to "H.M. SUBMARINE", then later to "H.M. SUBMARINES" which is still in use today.

It was not until 5 April 1958 that the Royal Navy approved a cloth badge with a stylized submarine for wear on the uniform left sleeve. Being both stylish and fashion-forward, the British submariners hated the new badge, promptly dubbing it "sausage on a stick". This badge was classified as a Trades badge and was worn by enlisted submariners only. There were three variations – the first (UK-1.1) was a gold submarine on black worn on the #1 uniform, dress square rig, the second (UK-1.2) was red thread on a black background for wear on the #2 uniform, the working square rig, and a third (UK-1.3) with a blue submarine on white background for the #6 uniform, tropical square rig. Since wearing the badge was optional, many sailors chose not to do so to avoid ridicule from other branches. The badges were available from private dealers and from Naval Supply (Stores) but were eventually dropped by the early 1960s due to lack of interest in wearing it.

UK-1.1

UK-1.2

UK-1.3

In the late 1960s, the UK always had at least one submarine stationed in Australia. So, a British boat and her crew was there when Australia issued their newly approved submarine insignia for a trial period. As noted in the Australian badge history, the crew of the HMS *Trump* was issued the Australian prototype badge by the RAN in 1968 to assess its use for a year. This created a rather awkward situation for the Royal Navy when the *Trump* returned to the UK with her crew sporting shiny new badges, making them the envy of the entire submarine fleet.

Acting relatively quickly, the British Admiralty fast-tracked the process for creating a new metal badge for British submariners, with the Royal Navy officially adopting a metal qualification insignia on 15 July 1971 in a ceremony where the badges were pinned on the majority of the submarine squadron. Based loosely upon the Australian badge, it is a black shield with a gold fouled anchor in the center, and St. Edward's crown with red enamel atop. These are supported by two inward-facing gold bottle-nosed dolphins. Unlike the Australian dolphins, these are dimpled, not smooth (UK-2). The first version issued, the W.J. Dingley dolphins, had a horizontal pin attachment so that it could be more easily pinned on during the initial pinning ceremony. The backs of some of the various manufacturers are shown (UK-2-WJD, UK-2-Firmin, UK-2-F&S, UK-2-TKS, UK-2-LB&B, UK-2-SM) with most badges being unmarked.

UK-2-WJD

UK-2r-WJD

UK-2-Firmin

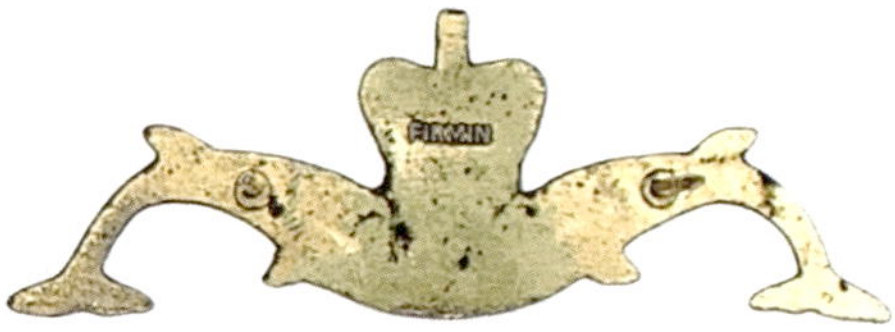
UK-2r-Firmin

UK-2-F&S

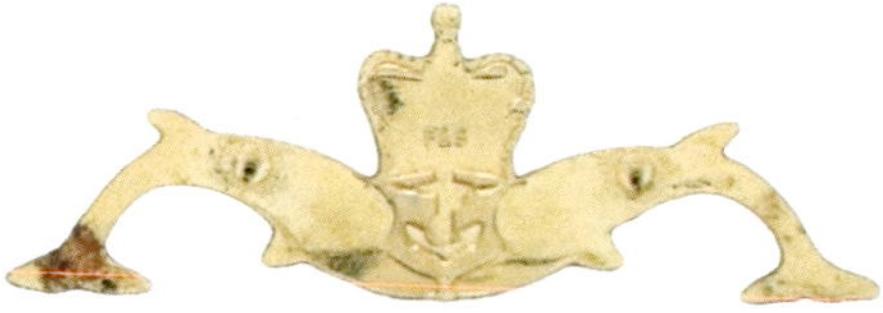
UK-2r-F&S

UK-2-TKS

UK-2r-TKS

UK-2-LB&B

UK-2r-LB&B

UK-2-SM

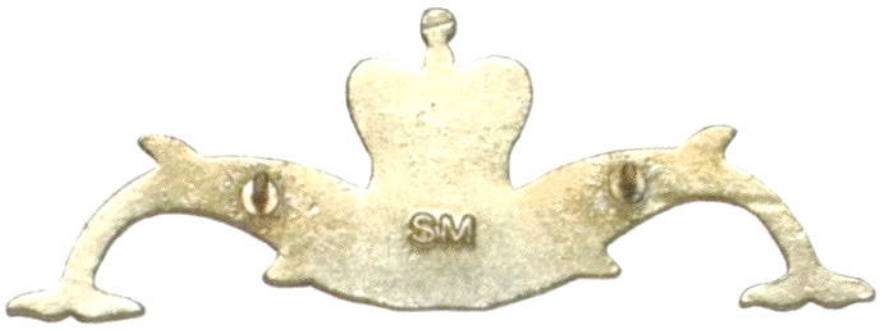

UK-2r-SM

To earn the "Kissing Kippers", as they became known, British submariners complete a three-part qualification process that includes basic submarine school, specialized trade training, and finally the on-board systems practical training. This process takes approximately six months to a year, depending on the trade. When earned, the badge is worn by all ranks on the left breast, above the medal bar, if applicable. These have remained the same basic design to the current date. There are subtle differences in attachment method due to manufacturer, but the front is held to strict deviation standards.

A slightly smaller mess dress version exists that is authorized for wear with miniature medals (UK-2m).

UK-2m

UK-2mr

There is also a black rectangular patch with gold dolphins that is authorized for use on work uniforms (UK-2C). A Royal Navy Submarines Blazer patch that is contoured in the shape of the badge exists but is not authorized for official use on uniforms (UK-UA-BLZR).

UK-2C

UK-UA-BLZR

A Trainee badge was approved for use in 2020 and is presented to personnel that declare for submarine service for the duration of their training until becoming fully qualified, when it is replaced by the official qualification insignia. This badge was first presented to an entire Training class on 25 September 2020 and is the same size and has the same design as the UK-2 but is completely mat black in color (UK-3).

UK-3

The UK is one of the few countries with Fleet Ballistic Missile (FBM) submarines, their first patrols conducted in April of 1969, but without a representative badge for 40 years. When finally issued in May 2009, pewter "Bomber" pins initially represented an individual completing between 1 and 19 tours of 30 days or more (UK-4.1), then a gold version was presented for 20+ patrols (UK-4.2). The official name of this badge is the "Continuous At Sea Deterrent" (CASD) badge. It is a profile of a British FBM submarine with an emerging missile surrounded by an atomic symbol in the middle with "ALWAYS READY" underneath. In conjunction with the 50th anniversary of the CASD service in April 2019, a silver version was created to represent 10-19 patrols completed (UK-4.3), with the pewter badge requirement shifting to 1-9 tours, and gold requirement remaining the same. Miniature mess dress versions exist for all three badges (UK-4.1m, UK-4.2m, & UK-4.3m). These badges are worn directly under the UK submarine qualification insignia on the left breast.

UK-4.1

UK-4.1m

UK-4.2

UK-4.2m

UK-4.3

UK-4.3m

UNITED STATES OF AMERICA (USN)

With a submarine history dating back to 1776 when David Bushnell's *Turtle* supposedly attempted an attack on the British line ship HMS *Eagle* in New York harbor, the United States has one of the longest histories of submarine use and development. This includes the submarine developed for the Confederate Navy, the *H. L. Hunley*, which became the first submarine to have a verified sinking of an enemy vessel sending the Union ship USS *Housatonic* to the bottom in 1864.

This history was built upon when John Philip Holland started developing submarines in the 1870s and by 1896 had developed the *Holland VI*. The U.S. officially joined the ranks of submarine operating countries when the *Holland VI* was purchased by the U.S. Navy 11 April 1900, commissioning it as the USS *Holland* (SS-01) on 12 October 1900. However, it would be over 20 years before the U.S. would join the ranks of countries with a submarine insignia.

The creation of the United States submarine qualification insignia was the culmination of an extended effort by Captain Earnest J. King, Commander of Submarine Division Three, who initially proposed a "distinguishing badge" for qualified submarine personnel in a letter to the Secretary of the Navy in a letter dated 13 June 1923. This was justified by the Army Air Corps already having a distinctive badge. Multiple designs were considered over the following year, but it was finally a design submitted by a jeweler, Bailey Banks and Biddle (BB&B), that was approved for wear by acting Secretary of the Navy Theodore Roosevelt Jr. on 29 March 1924, after it was recommended by the Chief of the Bureau of Navigation in a letter dated 24 March 1924.

BB&B worked in collaboration with and built upon a design by William C. Eddy who had created the class crest for the U.S. Naval Academy class of 1926 in 1922. The accepted submarine badge design shows a slightly port-facing bow-on submarine profile plowing through waves with bow planes rigged for diving that is based on a photo of a U.S. *O*-class submarine (PHOTO-US-O), with two inward-facing dolphins flanking the submarine, atop the bow planes. This basic design has remained unchanged to the present day making it one of the longest lasting designs in the annals of submarine insignia. The badge was approved

PHOTO-US-O
"Our Navy Collection - USNI Archives."

in gilt metal (USN-1) and in gold bullion (USN-1C) for officers, to be worn on the upper left chest above the ribbon bar. Multiple variations exist of the officer bullion badge, including some custom-made versions.

USN-1

USN-1r

USN-1C.1

USN-1C.2

A miniature metal badge was approved for officers on 1 May 1929 for use when miniature medals were worn. This badge was an exact duplicate of the USN-1, only half size (USN-1m).

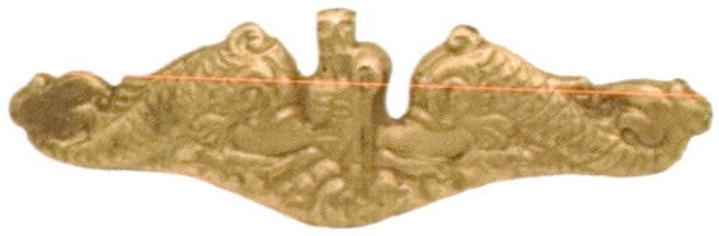

USN-1m

When the officer's badge was approved in 1924, a cloth submarine insignia was approved for enlisted personnel. It was originally issued in two versions, blue thread on white background (USN-2.1) for summer whites, and in white thread on black background (USN-2.2) for winter blues, to be worn on the right sleeve between the wrist and elbow for all enlisted personnel, including chiefs. The location was changed in 1947 to the upper left chest. Because of different manufacturers, there are literally dozens of different variations for the USN-2.1 and 2.2 cloth insignia, including variants with the submarine facing to starboard (USN-2.1S and USN-2.2S).

USN-2.1

USN-2.1S

USN-2.2

USN-2.2S

With the approval of khaki uniforms for officer and chief submariners in the 1930s, matching khaki cloth insignias were issued for wear by both (USN-2.3 & USN-2.4), though the chiefs still wore them on their sleeve and the officers on the chest.

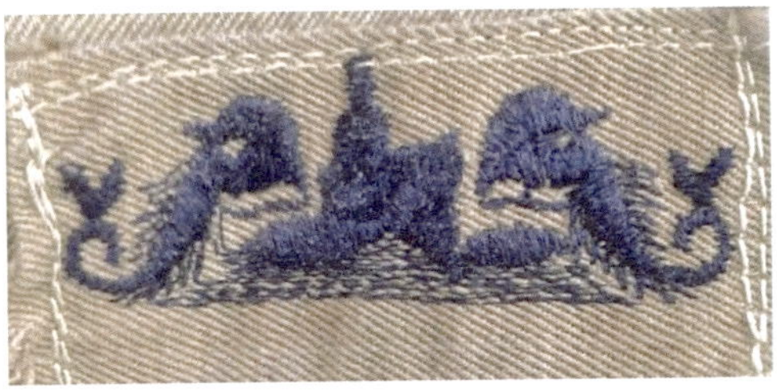

USN-2.3

USN-2.4

Silver bullion on dark blue cloth insignias were used by chiefs in the late 1930s and continued to be used after the war (USN-2.5 & USN-2.6).

USN-2.5

USN-2.6

Gray uniforms were in use for chiefs and officers briefly from 1943-1949. Cloth insignias using black thread on a gray background were in use for those uniforms (USN-2.7 & USN-2.8).

USN-2.7

USN-2.8

It was not until 21 September 1950 that a metal badge was approved for enlisted personnel. This badge is the same design as the officer's badge, but in silver metal instead of gold (USN-6 & USN-6m).

USN-6

USN-6m

While the official design for the U.S. Navy submarine qualification badge has remained unchanged for 100 years, that does not mean that certain artistic liberties have not been taken by various manufacturers. Over the years, several shades of gold and silver have been used as well as different finishes. Some manufacturers have deviated significantly enough from the original design that the variants they created acquired their own nicknames. Examples include: The "Deep Wave", first introduced by Hilborn & Hamburger in 1942 and later by Gemsco, with dolphins more closely resembling "Flipper", but still with scales (USN-1-HH-DW & USN-6-GEM-DW);

USN-1-HH-DW

USN-6-GEM-DW

The "Yokosuka" or "West Coast" dolphins introduced in Japan in the late 1950s (USN-1-YOK & USN-6-YOK);

USN-1-YOK

USN-6-YOK

The "SSN" badges introduced in the late 1970s (USN-1-SSN & USN-6-SSN);

USN-1-SSN

USN-6-SSN

And the "Littoral" dolphins (USN-1-LIT & USN-6-LIT) designed by Lee Lacey in the early 2000s.

USN-1-LIT

USN-6-LIT

While not all fall within the guidelines to be officially approved, all have been worn (or attempted to be worn!) as qualification badges by U.S. submariners. Some of these variants, especially the Deep Wave, were fabricated by multiple manufacturers, each adding their own touches, some of which we will address in the Manufacturers' section.

There are badges that appear to have been crafted as "Sweetheart" dolphins, some without rhinestones (USN-SWH.1) and some with (USN-SWH.2), made by an unknown manufacturer during WWII. There is no evidence that any of these were ever worn as approved dolphins.

USN-SWH.1

USN-SWH.2

One special badge that deserves mention was the brainchild of Admiral Robert English, the first Commander of Submarine Force Pacific Fleet during WWII. To reward commanding officers who had a particularly successful war patrol, he created what he called the "Roll of Honor" and presented each skipper meeting the entry criteria with a pair of gold dolphins having "ROLL OF HONOR" etched on the back, along with the skipper's name and the boat's name. This tradition was picked up and expanded upon by Admiral Charles Lockwood (known as "Uncle Charlie" to the submariners) who assumed command upon Admiral English's death in January 1943. There were around 90 of these dolphins created during the war and all known examples are H&H Deep Wave. The example shown here was presented to John H. Turner, skipper of the USS *Seal*, SS-183 for her 11th war patrol which earned him the Navy Cross (USN-1-ROH). These were often created as a set with a matching pair of mini dolphins (USN-1ROHm*). For a more detailed story on the Roll of Honor dolphins, see Dave Jones' excellent book, "*US Silent Service Dolphins and Combat Insignia 1924-1945*".

USN-1-ROH

USN-1r-ROH

A question that invariably arises and is subject to MUCH discussion and debate is: "Are the "dolphins" on the U.S. Navy submarine insignia supposed to be dolphins the mammal, or dolphin fish (mahi-mahi)?" The evidence is a little nuanced, but compelling. Stylized "dolphins" originated in Hellenic Greece and were used in English heraldry as early as the 13th century. Depictions of almost dragon-like fish were not unusual, and the examples almost always had scales and fins making them resemble fish. But these depictions with their many variations have been unequivocally proven by historians to represent bottle-nosed dolphins… without looking anything like them. Dolphins (the mammal) were considered to be the "king of all fish" in ancient times and were often stylized to make them appear more menacing and powerful. In the earlier part of the 20th century, the official name for the "fish" was "dolphin fish" or "dolphinfish". In the historical documents from Captain King and others associated with the development of the badge, the creatures gracing the badge are explicitly called "dolphins"… not "dolphinfish". Also, according to the U.S. Navy Undersea Museum, review of documents associated with Captain King's efforts to create an insignia shows that dolphins were "deemed a fitting symbol twice over: as the supposed patron of Sailors, and because they dive and surface similarly to submarines" and comments regarding "operating with intelligence." Obviously, references to mammalian dolphins and not the dolphin fish. So, despite the statements you will find on the Internet ("definitely mahi-mahi!"), in our eyes the evidence seems to point to Captain King intending for mammalian dolphins being the sea creatures depicted on U.S. submarine badges, no matter how heraldic or stylized they may be. However, opinions being what they are, the reader is left to form their own regarding this oh-so controversial topic.

U.S. Submarine Combat Patrol Insignia

It was not until 1942 that the U.S. Navy began to consider the possibility of a special insignia to identify successful submarine patrols. The first option considered was a small bronze numeral to be pinned above the qualification insignia for officers to coincide with the number of successful patrols, "1" for one, "2" for two, "3" for three and so on. For enlisted personnel, a small horizontal cloth stripe, 1" long x 1/8" wide was to be sewn above the submarine insignia on the sleeve. Next, a ribbon was considered that would be worn by both officers and enlisted personnel, with stars inserted to signify numbers of patrols. A third option considered was for officers to wear stars in holes drilled in their qualification insignia, with enlisted personnel still using the cloth stripes mentioned above. Fortunately, none of these options was approved by now Admiral King, and he ordered that a "Submarine Combat Patrol Insignia" be designed "similar to the recently authorized air crew insignia." So, like the submarine qualification badge, the submarine force was able to justify creation of a patrol badge based upon the air corps already having one.

Several designs were submitted, but the option selected was marked as the "Flying Fish Class." Designed by LT CDR Browder Gaines Nelson, the Submarine Combat Patrol Insignia (SCPI) was approved in a letter dated 2 April 1943 and was available to be issued 1 June 1943. It is authorized to be worn by any individual assigned to submarine duty on or after 7 December 1941 on a submarine that has completed a "successful" submarine wartime combat patrol during an officially declared war. This includes people not yet qualified on submarines. "Successful" was classified as either sinking or assisting in sinking an enemy vessel or completing a combat mission of comparable importance. The U.S. SCPI has not been issued to any active-duty personnel since the Korean War, yet new SCPIs have been in continuous production since after WWII and can still be purchased new today.

USN-3.1

USN-3.2

This badge is a silver starboard-facing WWII fleet *Gato*-class submarine profile riding the waves, a scroll with three holes centrally located at the bottom for stars to be installed. A gold star for single missions were originally authorized, then in October 1944, silver stars were approved to signify five missions. The badge itself represented the first successful patrol. Some examples had additional holes for stars drilled. This badge has two basic versions, the standard (USN-3.1) and the "Deep Wave" (USN-3.2), with the Deep Wave typically being somewhat shorter and having more layers of waves as the name implies. Miniatures of both styles exist (USN-3.1m & USN-3.2m) and were authorized for use when wearing miniature medals. The Submarine Combat Patrol Insignia was worn above the ribbon bar by enlisted personnel and just under the ribbon bar by officers.

USN-3.1m

USN-3.2m

This location was changed to under the ribbon bar for enlisted personnel after the metal enlisted version of the qualification badge was issued in 1950. Cloth insignias for this badge also exist for dress blues and khaki as shown here (USN-3C.1 & USN-3C.2 & 3C.3) with the qualification badges and ribbons also sewn on.

USN-3C.1

USN-3C.2

USN-3C.3

U.S. Submarine Medical Officer Badge

The U.S. submarine medical badge was the first individual "specialty" badge approved for submarine service and was authorized on 21 May 1943. It was issued to officers qualified as Submarine Medical Officers, a very small group of individuals with no more than 27 of them existing at any one time during the war. The majority of these personnel never served aboard submarines, however three are known to have been on a submarine patrol, with one being lost with the crew of the USS *Wahoo.*

Like the Combat Patrol badge, the initial proposals for the Medical Officer badge were much different than the final product. Because part of the qualification process required these personnel to qualify as divers, and because the field began with a focus on medical care for deep sea divers and expanded into submarine medicine, the original intent was to call the badge the insignia for "Medical Officers qualified in Submarine and Diving Medicine." This resulted in the initial proposed design being a Navy diving helmet flanked by two inward-facing heraldic dolphins (or "porpoises" as described in the documentation) (USN-4-PROPOSED SKETCH).

USN-4-PROPOSED SKETCH

The badge design that was ultimately selected is similar to the officer submarine qualification insignia, with an oval shield having a silver acorn and gold oak leaf of the medical branch replacing the submarine profile except for the bow planes. The original badges had a quite large silver acorn (USN-4.1) in the middle of most variants, and cloth insignias for this version are known (USN-4.1C). H&H also made one in the deep wave style (USN-4.2). The next version approved 11 August 1952 still featured the dolphins, but the bow planes were removed and replaced with waves. The oval shield remained the same but featured a noticeably smaller acorn (USN-4.3) and remains in use with some manufacturing differences to the current date. Miniatures of both versions are used for mess dress uniforms (USN-4.1m & USN-4.3m).

USN-4.1

USN-4.3

USN-4.1m

USN-4.3m

USN-4.2

USN-4.1C

U.S. Submarine Engineering Duty Officer Badge

The Submarine Engineering Duty Officer (EDO) badge, Type I, was approved for use 26 January 1950. EDOs qualifying for this badge need to serve for at least one year in a submarine-related waterfront assignment in a shipyard or as Supervisor of Shipbuilding Officer. This badge is not issued to a submarine's Engineering officer as is sometimes incorrectly assumed.

Type I was a Deep Wave version of the officer submarine qualification insignia, with a ¼" high gold capital "E" mounted center bottom on the waves (USN-5.1.1). No uniform standards can be found that show a style other than the Deep Wave badge for this version, but non-Deep Wave variants are known to exist (USN-5.1.2). Type I of the EDO badge was not in use for very long at all. The uniform regulations for Type II were issued on 11 August 1952 implementing the 1951 standards, superseding the previous design after less than two years in use. Type II has the two dolphins facing an inner circular design featuring a silver three-bladed propeller (USN-5.2). No examples of a Deep Wave version of this badge are known. This design remains in use to the current day, and a miniature version is known (USN-5.2m).

USN-5.1.1

USN-5.1.2

USN-5.2

USN-5.2m

Around 2015-2018, an effort was mounted to approve a badge for Limited Duty officers (LDOs) and Chief Warrant Officers (CWOs) qualifying as Repair Duty Officer (RDO) on submarine tenders and shore maintenance facilities. A badge was created by Vanguard with the exact same design as the Type II EDO badge, but with an antique brass finish (USN-5-UA-RDO). Though a small quantity of these badges was manufactured, the program was never finalized, and thus the badges were never approved for issue.

USN-5-UA-RDO

U.S. Submarine Supply Corps Badge

Officers of the U.S. Navy Supply Corps who qualify aboard submarines were rewarded with their own badge 7 February 1964. This badge is the standard submarine dolphins with a gold oak leaf symbol of the Supply Corps replacing the submarine profile (USN-7), and a miniature is also approved (USN-7m). While the "Chop" (called the Chop because the oak leaf resembles a pork chop) is not eligible for command, they can eventually qualify as Diving Officer of the Watch and other duty stations. This badge is unchanged since initial issue with minor differences between manufacturers.

USN-7

USN-7m

U.S. SSBN Deterrent Patrol Insignia

On 28 January 1969 the U.S. Navy authorized issuance of an SSBN Deterrent Patrol Insignia for officers and enlisted personnel completing strategic deterrent patrols aboard fleet ballistic missile submarines ("boomers"). Initially, awarding of the insignia was made retroactive to the first strategic patrol conducted by the USS *George Washington* on 27 January 1961. In 2004, this was modified to include those submariners on the five submarines conducting the first nuclear deterrent patrols while carrying Regulus missiles from 1959 to 1964.

USN-8C

This badge shows a starboard-facing *Lafayette*-class SSBN with an emerging Polaris missile superimposed on the front, encased in a nuclear symbol. A scroll is across the bottom of the badge with holes for stars signifying patrols, similar to the U.S. SCPI. This badge is issued in silver (pewter) for personnel completing between 1 and 19 patrols (USN-8.1). Beginning in 1979, personnel with 20 or more patrols are awarded a gold version of the same badge (USN-8.2), with the holder of the most deterrent patrols in the submarine force being presented with the Neptune Award. Miniatures of both are approved for use (USN-8.1m & USN-8.2m), and cloth insignias are also known (USN-8C).

USN-8.1

USN-8.2

USN-8.1m

USN-8.2m

U.S. Deep Submergence Badge

The U.S. Navy Deep Submergence program gathered steam after WWII, and finally came to fruition with the first deep submersible, the Bathyscaphe *Trieste,* launching in 1959. It was not until 1969 that an effort was initiated by Captain Don Walsh, the U.S. Navy's first deep submersible pilot, to implement a qualification badge for personnel assigned to the deep submersibles. Captain Walsh worked with a jeweler in Italy to have a prototype created that he submitted to the Navy for approval. This prototype was a solid gilt-colored badge featuring the *Trieste* in the center, flanked on the sides by two diving stylized heraldic dolphins, Neptune's trident emerging from the top center, and a scroll at the bottom with three holes (USN-DS-PTYP). The three holes were intended to be for stars that would signify crewmember (one star), co-pilot (two stars), and pilot (three stars). Only three of the original prototype badges were created, with one retained by Captain Walsh, and two submitted to the U.S. Navy for the approval process. High quality, but unauthorized hollow back copies of this badge have been made in gilt and silver metal, but these have no official standing (USN-DS-UA1 and USN-DS-UA2). Some of these were possibly used by the crews of the submersibles in the 1970s despite not being authorized.

USN-DS-PTYP

USN-DS-UA1

USN-DS-UA2

There was quite a gap following Captain Walsh's initial submission before the U.S. Navy finally approved an official Deep Submergence qualification badge on 6 April 1981. Borrowing liberally from the originally submitted design, the badge has the *Trieste* in the center, flanked by two diving heraldic dolphins similar to those featured on the submarine qualification insignia, and the fork of Neptune's trident emerging top center. The Navy decided against the scroll and stars of the original submission. This badge is issued in gold for officers (USN-9) and silver for enlisted personnel (USN-10). This is the last official U.S. submarine badge issued. It is worn on the left chest, under the ribbon bar, and remains unchanged to the present day.

USN-9

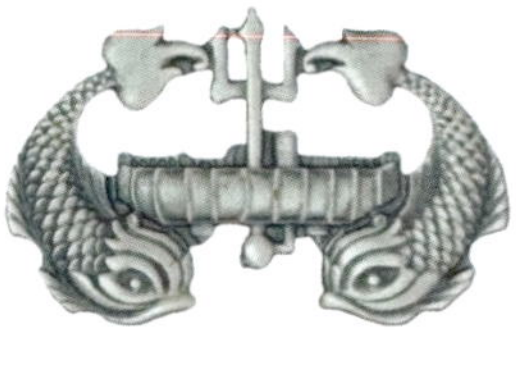

USN-10

Deep Submergence Rescue Vehicle Badge

The deep submergence rescue vehicles (DSRVs) *Mystic* and *Avalon* were both launched in 1971 as a response to the tragic loss of the USS *Thresher*. Though both were in service until the 2000s, there was never an authorized DSRV badge issued, possibly because personnel operating the DSRVs qualified for the U.S. Navy Deep Submergence badge. Prototypes were created by Robbins, but never approved. The design was a port-facing DSRV with the transfer skirt center bottom. And yes, that is a transfer skirt on the bottom of the DSRV, and not a sail on the top as is often erroneously thought. These were made in gold and silver (USN-DSRV-UA.1 & USN-DSRV-UA.2) with some badges having the Robbins maker's mark on the back, however, many more versions have been made having no maker's mark. None of these variations have any official standing.

USN-DSRV-UA.1

USN-DSRV-UA.2

Diesel Boats Forever Pin

The Diesel Boats Forever (DBF) pin was created in 1967 as a bit of a stick in the eye to Admiral Rickover's nuclear submarine community. After a discussion aboard the USS *Barbel*, SS-580, regarding the superiority of diesel boats due to the perceived unreliability of the nuke boats, the crew held a competition to create a diesel boat-specific badge. ETR3(SS) Leon Figuerito submitted the winning sketch (USN-DBF-SKTCH), and when the boat reached Yokosuka, Japan, crew members took the sketch into town and had badges made.

The badge is a port-facing fleet diesel submarine on the surface with "SS" on the sail, two mermaids laying on top of the submarine and clasping hands in front of the sail, a scroll centered at the bottom with "DBF" in the middle, and holes for stars on either side (USN-DBF). While many assume the stars applied were to represent the number of diesel boats served on, it was confirmed by those involved in the creation of the badge that the stars were to intended represent each time a diesel boat being served on had to "get underway for a broke-down nuke", though often the stars ended up being added for particular missions. While the last of the U.S. diesel boats have long been decommissioned, this badge is still proudly worn by many U.S. submarine veterans at various functions.

USN-DBF-SKETCH

USN-DBF1

USN-DBF2

Maryland Naval Militia

The Maryland Naval Militia (MNM) was an organization that had been in place since 1774. In the early 1970s, though in its waning years, there was a resurgence in interest in developing MNM-specific rank, rating and specialty insignias. One of the items designed during this time was a submarine badge for MNM members who had been qualified aboard submarines. It was a standard U.S. submarine qualification badge, with the Maryland state seal mounted in the center instead of the bow of a submarine. This was made in gold for officers (USN-MNM-1) and silver for enlisted personnel (USN-MNM-2).

It is not known how many badges were made in the early 1970s before the MNM became defunct in 1975. These badges were made locally in Maryland and had no maker's marks. However, many copies or reproductions exist, and it can be very difficult, if not impossible, to tell the difference between genuine period pieces and reproductions.

USN-MNM-1

USN-MNM-2

Subdued Badges

With a history that is somewhat shrouded in mystery, the authors have been unable to specifically identify exactly when black, or "subdued" badges first appeared on the scene for submariners. Sources indicate that they may have been used by personnel making cross-service moves as early as the 1970s when transferring to a service using olive drab uniforms. Use appears to have increased after implementation of Battle Dress Uniforms (BDUs) in1981. Prior to the official manufactured versions, personnel qualified on submarines who transferred to billets authorized to wear either the olive drab or BDUs would blacken standard badges by various methods to emulate personnel with other qualification insignias in the unit. When BDUs were approved for submarine personnel in the 1990s, a full range of submarine insignias was created (USN-SUBDUED). The badges shown here were manufactured by Vanguard, but other manufacturers exist.

USN-1-SUBDUED

USN-4-SUBDUED

USN-5-SUBDUED

USN-7-SUBDUED

USN-8-SUBDUED

USN-3-SUBDUED

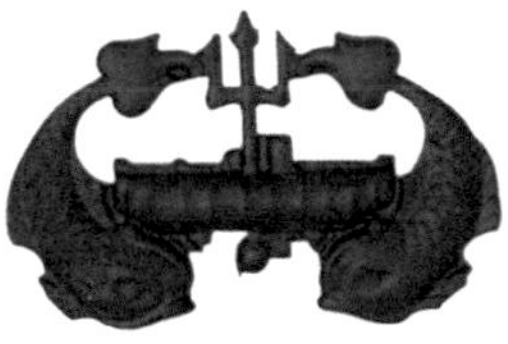

USN-9-SUBDUED

Individual Unit Badges

While multiple other countries have boat badges or pins that are specific to the submarine, most are never worn as an official part of the uniform. U.S. submarine individual unit badges are customized, stylized qualification badges that are manufactured for the members of a specific submarine, typically weaving characteristics the crew members feel represent their boat into the design. In most cases, these badges are actually allowed to be worn by crew members in place of their official qualification insignia for ship functions. In some cases, unauthorized copies have been made of these boat badges and sold by individuals. These badges were typically created for and worn by crew members only and are sometimes sold by the boat to make money for the recreation fund. The unauthorized copies can infringe upon the ability of the crew members of active boats to raise funds.

The following are examples of such individual unit badges that are in as close to chronological order as we could ascertain:

The USS *Parche* created the first example of an individual unit badge that we could find. Though details are slim, this badge was apparently the idea of the CO at the time, Captain Buchanan. This is a quite rough looking badge, similar in all ways to the official submarine qualification badge except the *O*-class submarine bow has been replaced with a slightly port-facing profile of a round-hulled *Sturgeon*-class submarine complete with fairwater planes (USN-PAR). These were issued to personnel qualifying on the *Parche* around 1985, and it is unknown if there were also officer dolphins created in this same style as only enlisted dolphins have been identified by the authors to date.

USN-PAR

The next such example was issued in 1989 for the commissioning of the USS *Pennsylvania*. Being a state strongly identified with coal, each submariner that was part of the commissioning crew received a pair of anthracite dolphins made using the Gemsco Deep Wave design (USN-PEN). These had a horizontal pin glued to the back for an attachment point.

USN-PEN

The USS *Philadelphia* Reunion or "688" badges (USN-688-OFF & USN-688-ENL) were designed by a former crew member, Guy Hartsough, for the boat's reunion in 2002 in an effort to have a badge more representative of the 688-class boats with a bit of a 3D look. Attempts were made to have them authorized as official dolphins by BUPERS for the Navy Uniform Regulations, however they were not approved. Modern reproductions have been made.

USN-688-OFF

USN-688-ENL

USS *Seawolf* dolphins simply replaced the *O*-class submarine profile with a *Seawolf*-class profile (USN-SW-OFF & USN-SW-ENL). These badges appeared in the mid-2000s and were made in gold and silver by the badge manufacturer, Krew. There was a bit of an attempt to have them replace the original badge for the *Seawolf*-class submariners, but that never materialized.

USN-SW-OFF

USN-SW-ENL

USS *Columbia* Tiger Sharks designed by Shane Petitto in 2013 feature a bow-on view of a *Los Angeles*-class submarine profile plowing through waves, flanked by two inward-facing tiger sharks. Issued in gold for officers and silver/gray for enlisted personnel, the etching to give the tiger sharks their distinctive stripes is particularly vivid on the silver version due to the dark contrast used in the lines (USN-COL-OFF & USN-COL-ENL).

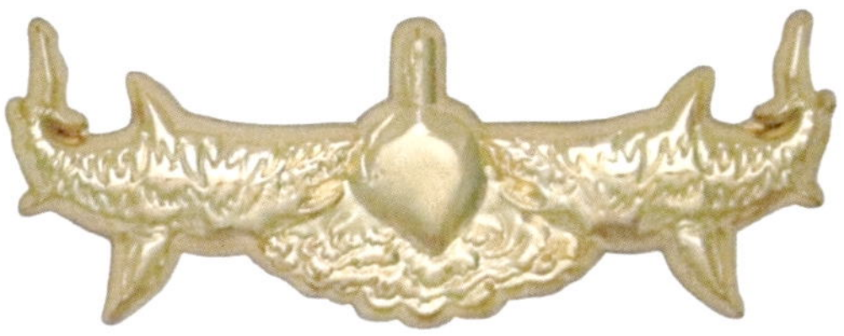

USN-COL-OFF

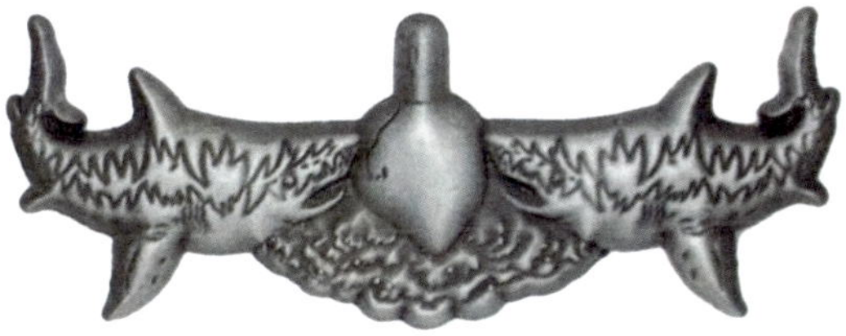

USN-COL-ENL

Borrowing a page from the USS *Pennsylvania*, the USS *Scranton's* crew had anthracite coal dolphins crafted in 2015 for the boat's 25th anniversary. Originally planned to be based on the H&H Deep Wave dolphin design, the prototypes did not work out, so they also ended up also using the Gemsco design for the badges. These can be differentiated from the *Pennsylvania's* badges by the fact that they are quite a bit thinner and were made with a pierced die (USN-SCRN). These were created by a former diesel boat submariner, William "Shut" Klose using his original qualification dolphins for the mold. "Shut" was a former enlisted sailor who retired as a Lieutenant and was a driving force behind several badges/pins including the Electric Boat *Virginia*-class pins.

USN-SCRN

Several years later, the *Scranton* crew went a step further and created their Steampunk Dolphins. A slightly port-facing profile of a locomotive belching smoke, flanked by two steampunk stylized dolphins, they were issued in gold for officers and silver for enlisted personnel (USN-SCRN-SP-OFF & USN-SCRN-SP-ENL). These are some of the newer boat badges issued.

USN-SCRN-SP-OFF

USN-SCRN-SP-ENL

Taking a full port turn and channeling their ship's motto, "Gator Done!", the USS *Mississippi* crew created gold and silver badges with alligators replacing the dolphins of the standard qualification insignia in 2018 (USN-MISS-OFF & USN-MISS-ENL). Unfortunately, this is one of the most-copied boat badges.

USN-MISS-OFF

USN-MISS-ENL

In 2021, inspired by the ship's motto, "Give 'em Hel!", LCDR Jon Luetkenhoelter the weapon's officer of the USS *Helena* at the time, designed and created *Helena's* "Helfish". Metal badges in gold and silver featuring the demonically stylized Helfish flanking a bow-on view of a *Los Angeles*-class submarine profile with fairwater planes were given to each crew member and are presented to all newly qualified personnel (USN-HEL-OFF & USN-HEL-ENL). Cloth patches were also distributed (USN-HEL.C).

US-HELENA-OFF

US-HELENA-ENL

USN-1-HEL.C

The USS *Ashville* Ghost dolphins were developed in the early 2020s, replacing the standard heraldic dolphins with skeletal fish as a nod to the city's famous ghost tours and the boat's nickname as the "Ghost of the Coast." A head-on profile of a *Los Angeles*-class boat with bow planes and a line of skulls at the bottom complete this macabre and interesting rendition of the U.S dolphins (USN-ASH-OFF & USN-ASH-ENL). These are issued to crew members in gold and silver versions and are also among the most heavily copied boat badges.

USN-ASH-OFF

USN-ASH-ENL

The USS *New Jersey* crew created a pair of red dolphins called the Devil Fish around 2021 that they use for crew events. As a connection to the state of New Jersey and the so-called "Jersey Devil" they use the red color to identify themselves as the "Devils of the Deep." These red dolphins are worn by both officers and enlisted personnel (USN-NJ).
They have a single pair of powder-coated red dolphins that they use for every individual qualifying on the boat. The intent is to use them for the life of the ship so that the very first and very last sailor will have been pinned with the same pair, linking all members of the crew.

USN-NJ

Borrowing liberally from the University of Texas Longhorns' mascot, the USS *Texas* "Texas Fish" were designed by FT1(SS) Paul Egan and implemented in 2023. Issued in both gold and silver, a longhorn skull replaces the submarine profile in the center (USN-TEX-OFF & USN-TEX-ENL).

USN-TEX-OFF

USN-TEX-ENL

After a long hiatus, the USS *Pennsylvania* crew jumped back into designing an individual unit badge with the creation their Liberty Bell dolphins. This badge has the dolphins flanking the historic bell, with "PASS AND STOW", "PHILADA", and "MDCCLIII" all clearly visible beside the famous crack, but also "735" to represent the hull number. Gold for officers and silver for enlisted (USN-PEN.1-OFF & USN-PEN.1-ENL), the crew took a unique approach to attachment methods creating pins that have a magnetic clasp on the back.

USN-PEN.1-OFF

USN-PEN.1-ENL

The USS *Vermont* crew took inspiration from the local legend of the Lake Champlain monster, "Champ" to create a badge. Designed by FT2 Beasley and FT2 Morales, the "dueling Champies" badge features a port-angling *Virginia*-class submarine profile in front of a profile of the state of Vermont, flanked by two serpentine lake monsters, the left one with a green eye and the right with a red eye, emulating navigation running lights (USN-VER). Only silver badges were made before COMSUBLANT issued an official message banning individual unit badges. We hope this ban does not last, and that the tradition of these individual unit badges can continue. We look forward to featuring any new additions in a future edition.

USN-VER

U.S. Submarine Qualification Badges by Manufacturer

The purpose of this section is simply to show some of the badges made by the various manufacturers, and to also feature many of the different maker's marks on the backs used by manufacturers over the years. In general, this is a pictorial section, with the companies listed in alphabetical order. This will NOT include ALL manufacturers, as many simply used the design/mold from other manufacturers. We will not highlight the design differences nor similarities between the many badges, leaving that to the reader to ascertain. We will also not dive into the whole "WWII or earlier" or "post-WWII" discussion, as there is no 100% accurate way to make that assessment other than provenance and experience with some characteristics listed in other books or on various websites.

Amcraft of Attleboro, Mass

USN-1-AMCRAFT.1

USN-1-AMCRAFT.1r

American Insignia Company, NY
AMICO

USN-1-AMICO.1

USN-1-AMICO.1r

USN-1-AMICO.2r

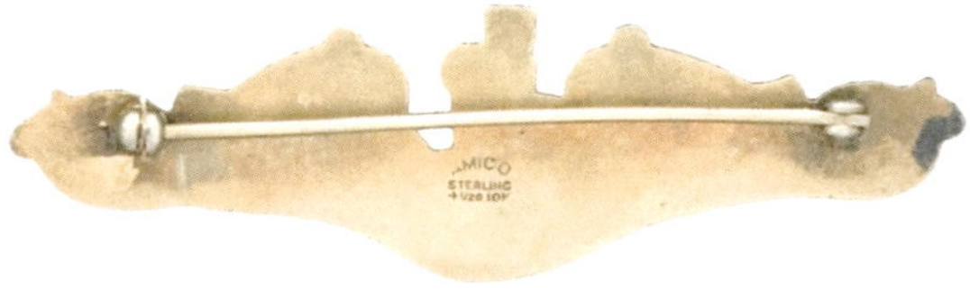

USN-1-AMICO.3r

USN-1-AMICO.4r

USN-1-AMICO.5r

USN-4-AMICO.1

USN-4-AMICO.1r

USN-3-AMICO.1

USN-3.AMICO.1r

USN-6-AMICO.1

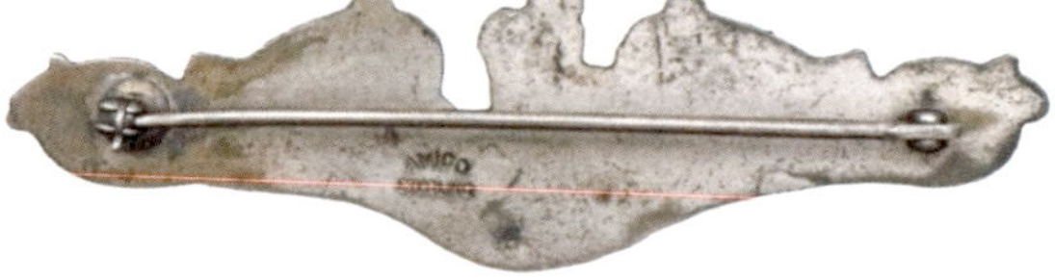

USN-6-AMICO.1r

USN-6-AMICO.2r

L.G. Balfour, Attleboro, MA
(LGB, B21).

USN-1-LBG.1

USN-1-LGB.1r

USN-1-LGB.2r

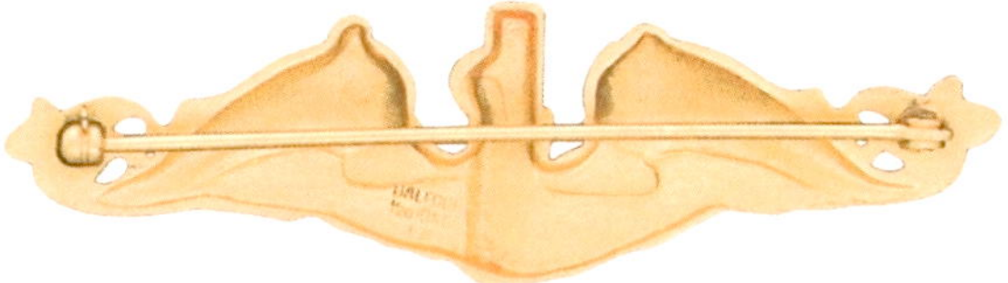

USN-1-LGB.3r

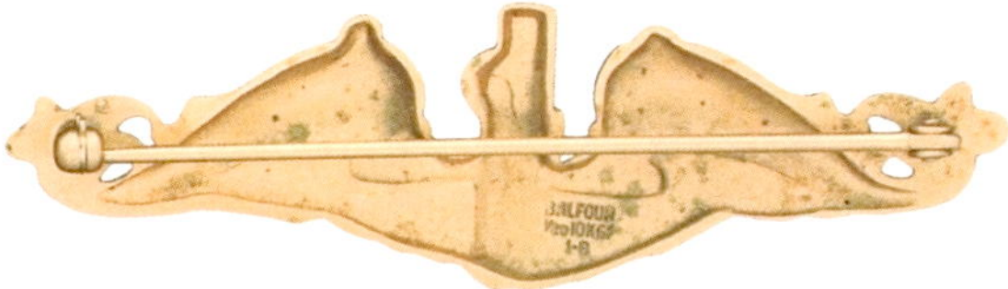

USN-1-LGB.4r

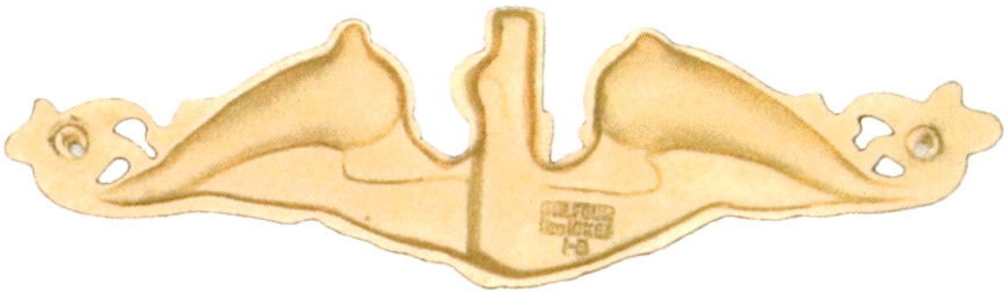

USN-1-LGB.5r

USN-1-LGB.6r

USN-6-LGB.1

USN-6-LGB.1r

USN-6-LGB.2r

USN-6-LGB.3r

USN-6-LGB.4r

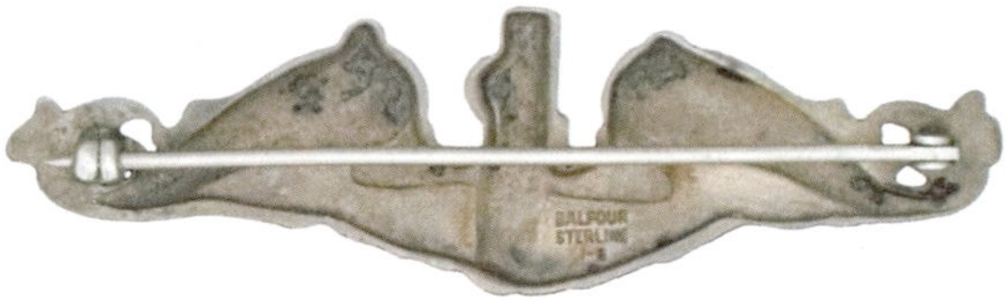

USN-6-LGB.5r

USN-6-LGB.6r

USN-6-LGB.7r

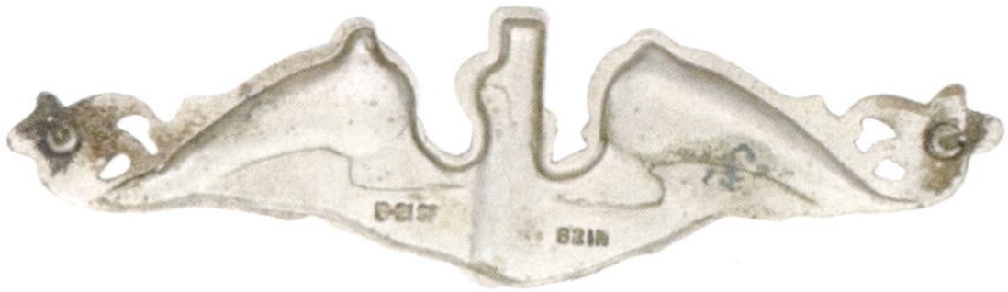

USN-6-LGB.8r

Wallace Bishop
Australia

USN-1-WB.1

USN-1-WB.1r

V.H. Blackinton Company, Attleboro Falls, MA
(VHB, B24)

USN-1-VHB.1

USN-1-VHB.1r

USN-3-VHB.1

USN-3-VHB.1r

USN-4-VHB.1

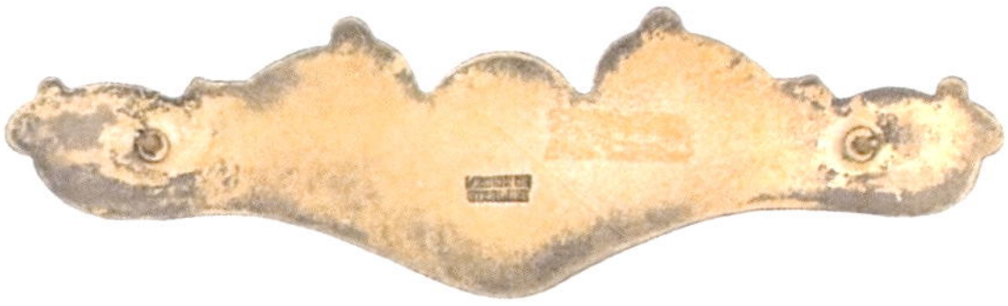

USN-4-VHB.1r

USN-4-VHB.2

USN-4-VHB.2r

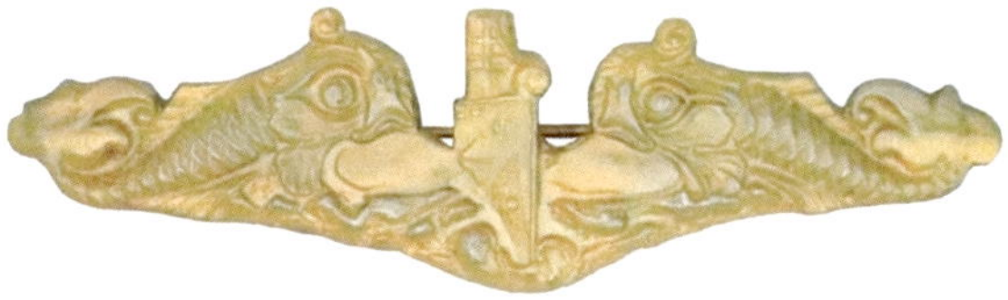

USN-1-VHB.2

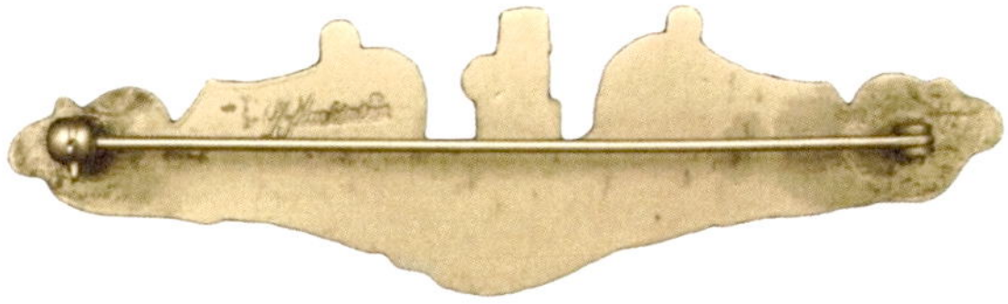

USN-1-VHB.2r

USN-4-VHB.3

USN-4-VHB.3r

USN-6-VHB.1

USN-6-VHB.1r

Denmark's Military Equipment Corp., New York/Astoria, NY.
(6D, D22)

USN-6-DEN.1

USN-1-DEN.1

USN-1-DEN.1r

USN-6-DEN.2

USN-6-DEN.2r

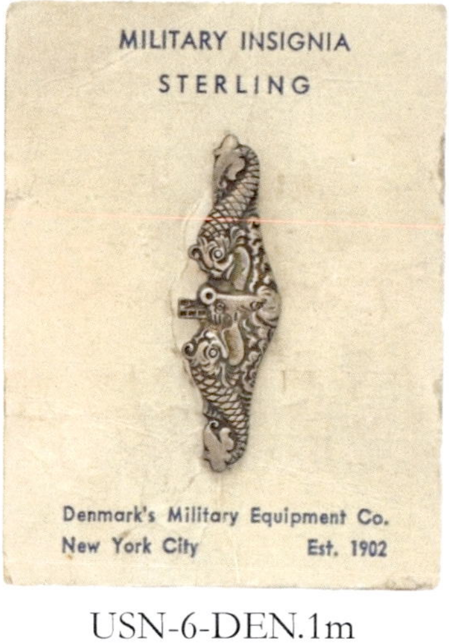

USN-6-DEN.1m

Gemsco (General Merchandising Co.) Milford, CT.
(G2, G22)

USN-1-GEM.1

USN-1-GEM.1r

USN-1-GEM.2.1

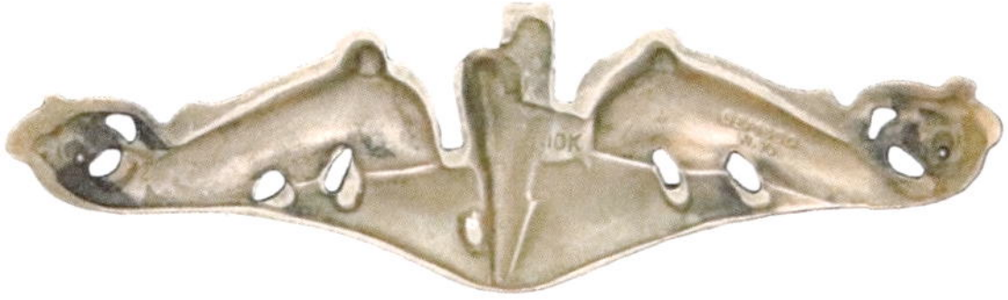

USN-1-GEM.2.1r

USN-1-GEM.2.2r

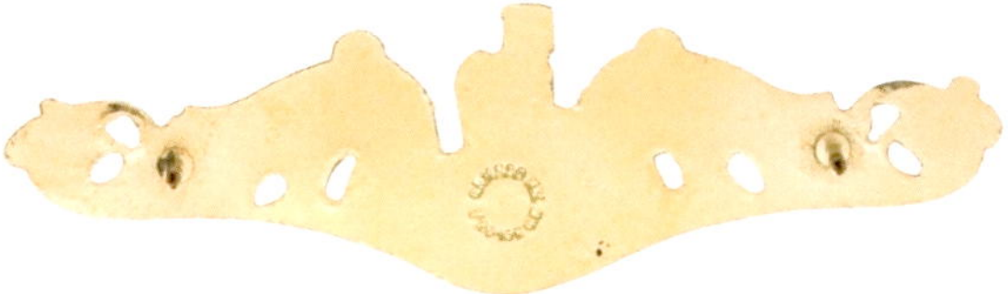

USN-1-GEM.2.3r

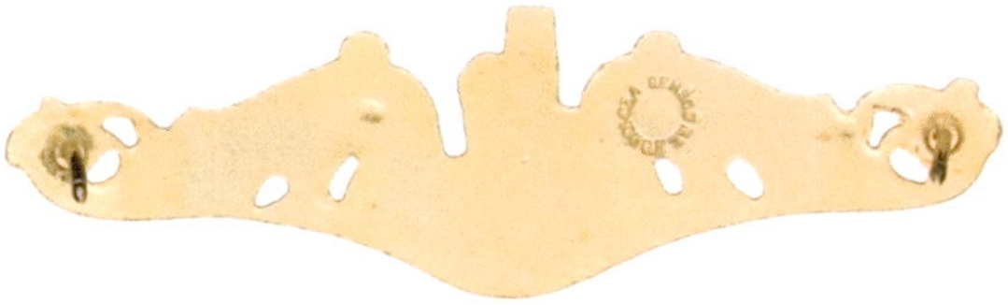

USN-1-GEM.2.4r

USN-1-GEM.3.1

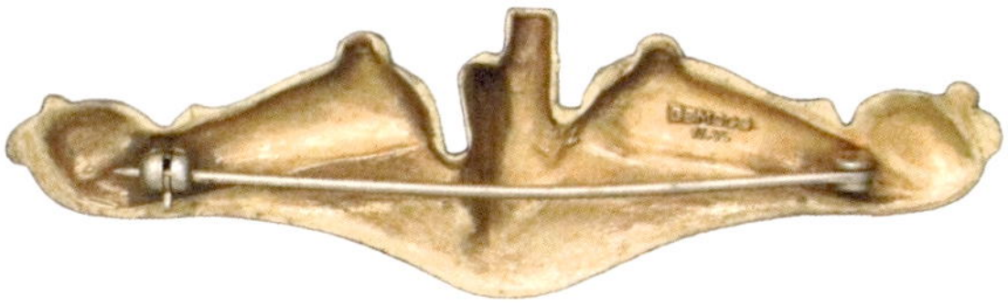

USN-1-GEM.3.1r

USN-3-GEM.1

USN-3-GEM.1r

USN-4-GEM.1.1

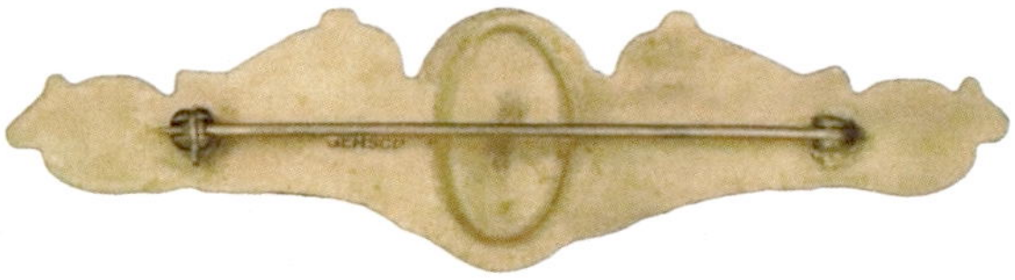
USN-4-GEM.1.1r

USN-4-GEM.2.1

USN-4-GEM.2.1r

USN-4-GEM.2.2

USN-4-GEM.2.2r

USN-6-GEM.1.1

USN-6-GEM-1.1r

USN-6-GEM.2.1

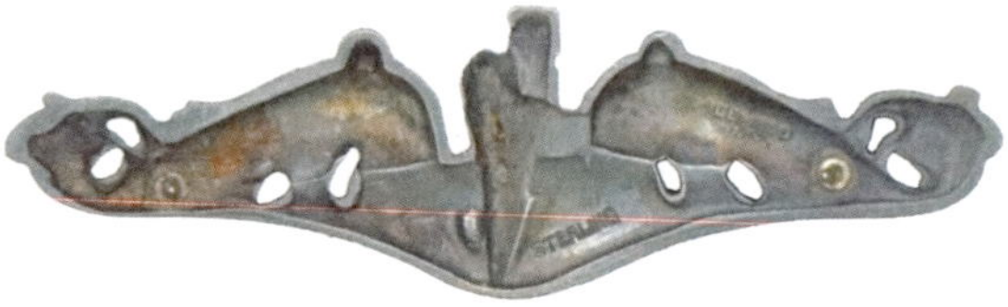
USN-6-GEM.2.1r

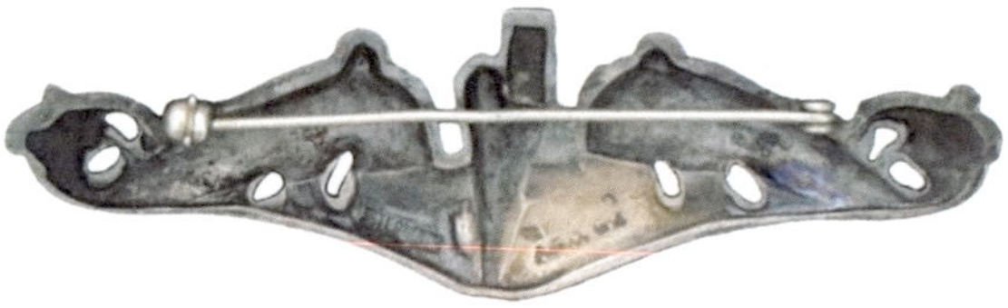
USN-6-GEM.2.2r

USN-6-GEM.2.3r

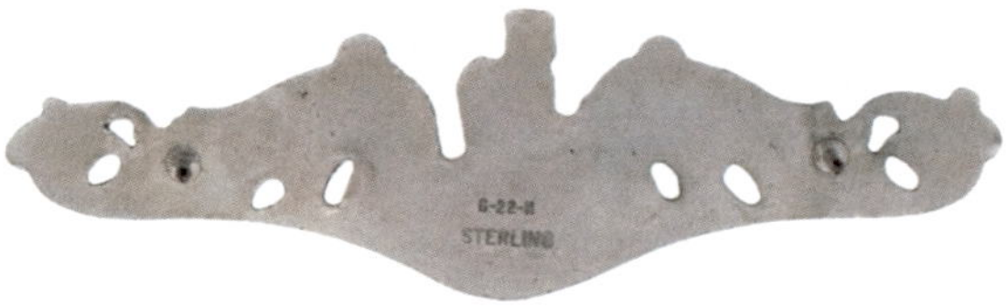

USN-6-GEM.2.4r

USN-6-GEM.3

USN-6-GEM.3r

USN-7-GEM.1

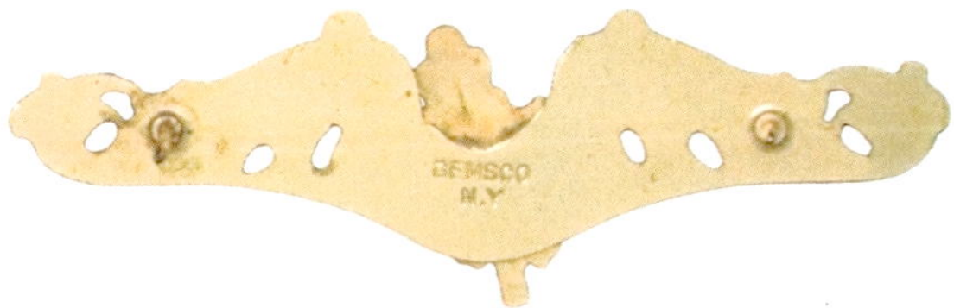

USN-7-GEM.1r

USN-8.1-GEM.1

USN-8.1-GEM.1r

USN-8.1-GEM.2r

Gull Associates Inc
Old Lyme, CT

USN-1-GULL.1

USN-1-GULL.1r

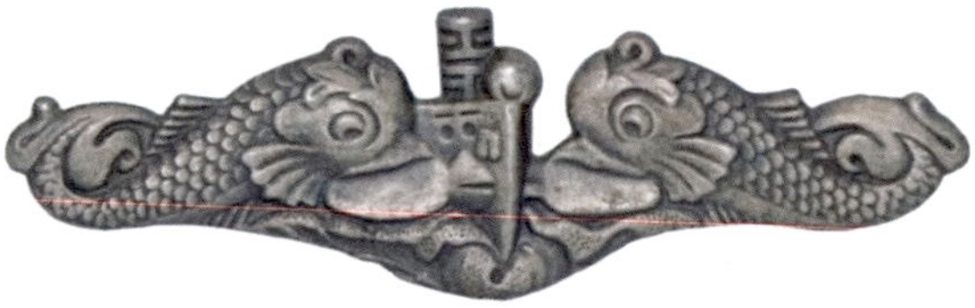

USN-6-GULL.1

USN-6-GULL.1r

Hilborn-Hamburger, New York, NY
(H-H/eagle logo, H-H Imperial, H24)

USN-1-HH.1.1

USN-1-HH1.1r

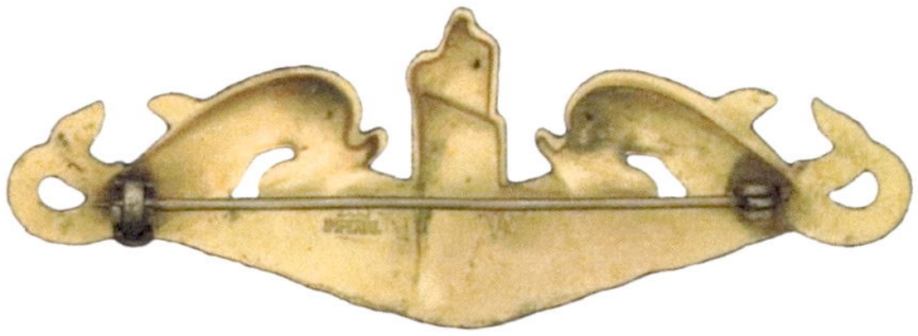

USN-1-HH-DEEP-1.2r

USN-1-HH-DEEP-1.3r

USN-1-HH.1.4r

USN-1-HH.2

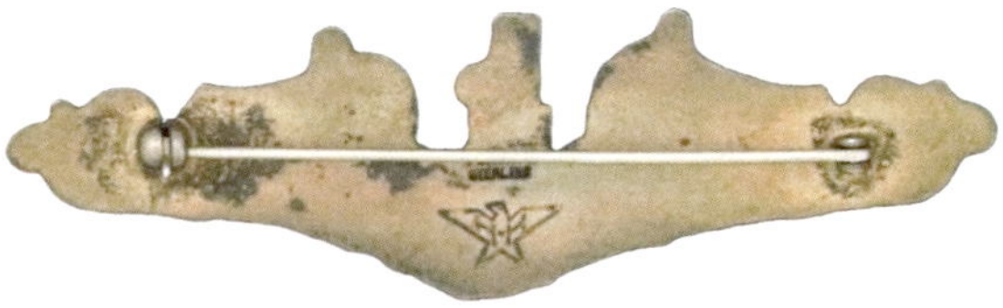

USN-1-HH.2r

USN-4-HH.1

USN-4-HH.1r

USN-3-HH.1.1

USN-3-HH.1.1r

USN-3-HH.1.2r

USN-3-HH.1.3r

USN-3-HH.2.1

USN-3-HH-2.1r

USN-6-HH.1

USN-6-HH.1r

USN-8.2-HH.1

USN-8.2-HH.1r

USN-8.1-HH.1

USN-8.1-HH-1r

USN-8.1-HH.2

USN-8.1-HH.2r

USN-8.1-HH.3

USN-8.1-HH.3r

JAPAN

USN-6-JAPAN.1

USN-6-JAPAN.1r

USN-6-JAPAN.2

USN-6-JAPAN.2r

Krew Inc.
Attleboro,MA. (1K, K21)

USN-5-KREW.1

USN-5-KREW.1r

N.S. Meyer
New York, NY. (9M, M22)

USN-1-MEYER.1.1

USN-1-MEYER.1.1r

USN-1-MEYER.1.2r

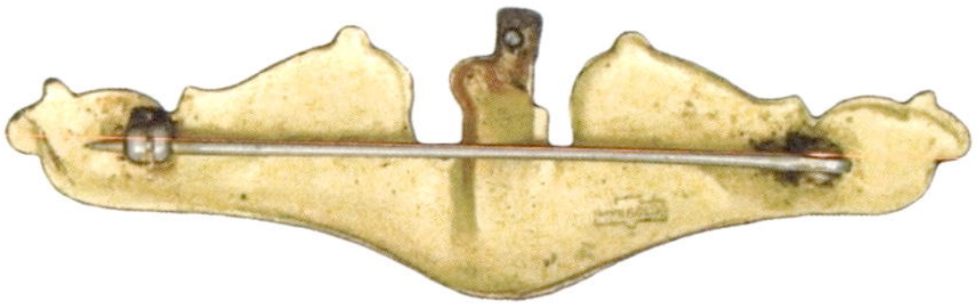

USN-1-MEYER.1.3r

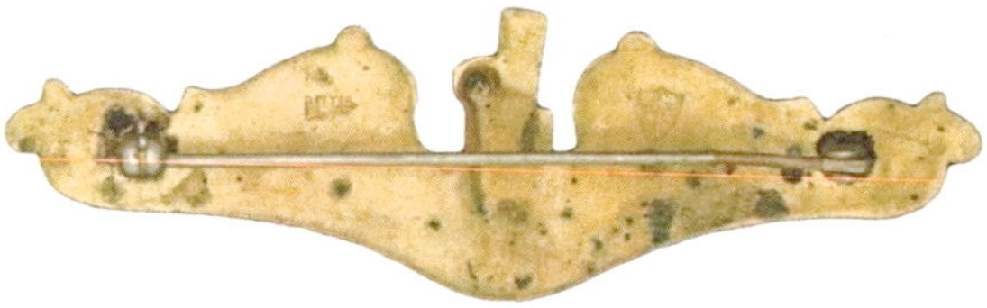

USN-1-MEYER.1.4r

USN-1-MEYER.2.1

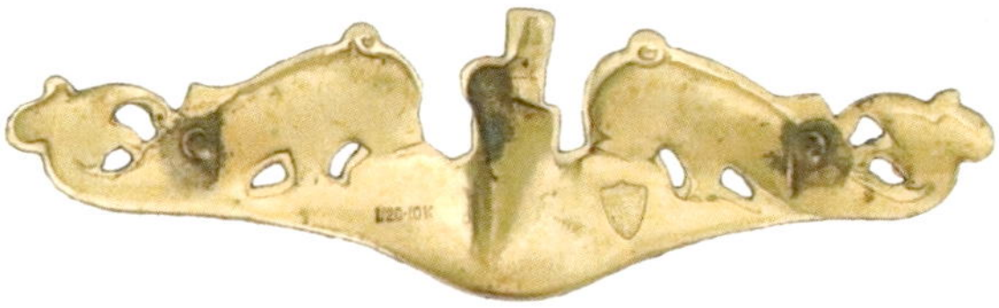

USN-1-MEYER.2.1r

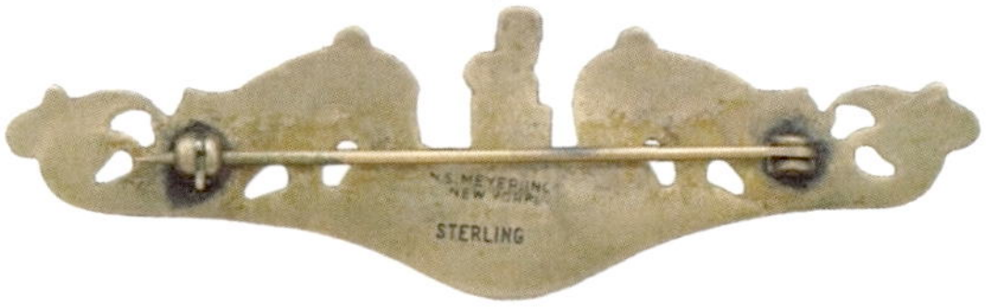

USN-1-MEYER.2.2r

USN-1-MEYER.2.3r

USN-3-MEYER.1

USN-3-MEYER.1r

USN-3-MEYER.2r

USN-3-MEYER.3r

USN-6-MEYER.1

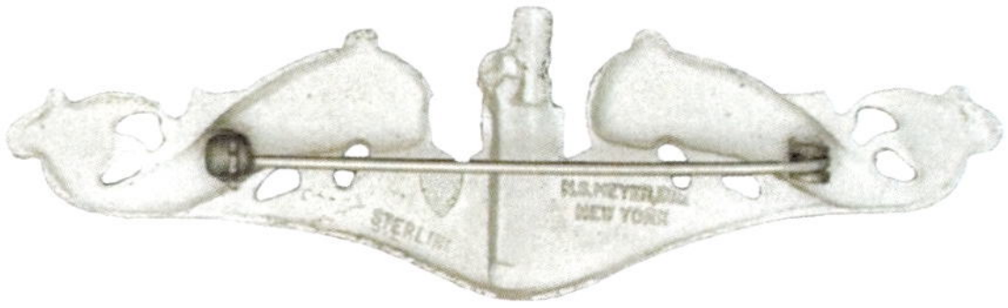

USN-6-MEYER.1r

USN-6-MEYER.2r

USN-7-MEYER.1

USN-7-MEYER.1r

USN-8.1-MEYER.1

USN-8.1-MEYER.1r

NANCO

USN-1-NANCO.1

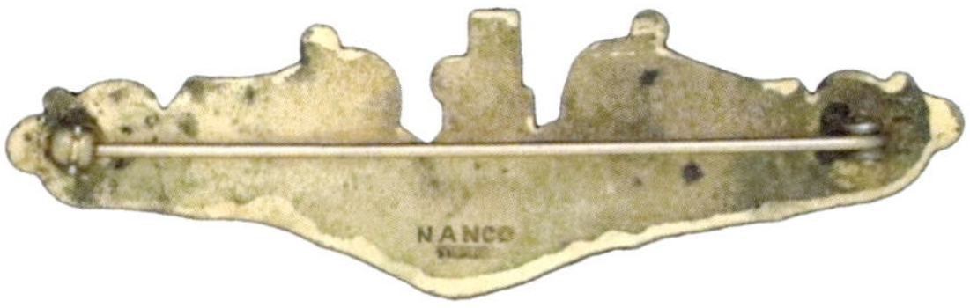

USN-1-NANCO.1r

Nose Dragon

USN-1-DRAGON.1

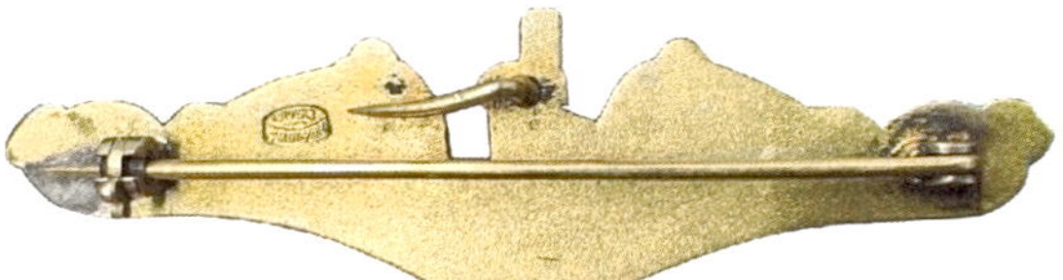

USN-1-DRAGON.1r

Officers Equipment Co.
Madison, NJ. (O22)

USN-1-OFFICERS.1

USN-1-OFFICERS.1r

USN-6-OFFICERS.1

USN-6-OFFICERS.1r

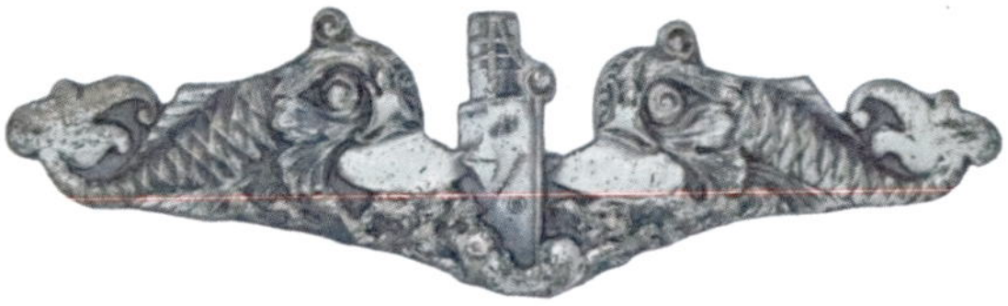
USN-6-OFFICERS.2

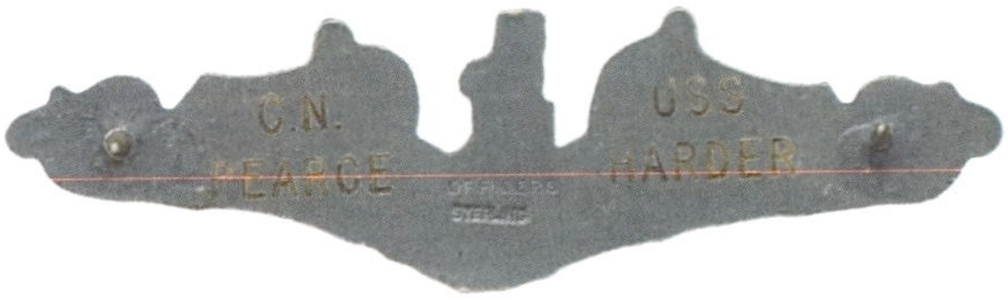

USN-6-OFFICERS.2r

Josiah Odence
MFG & Distributor of Military Insignia New York & San Diego

USN-1-JO.1

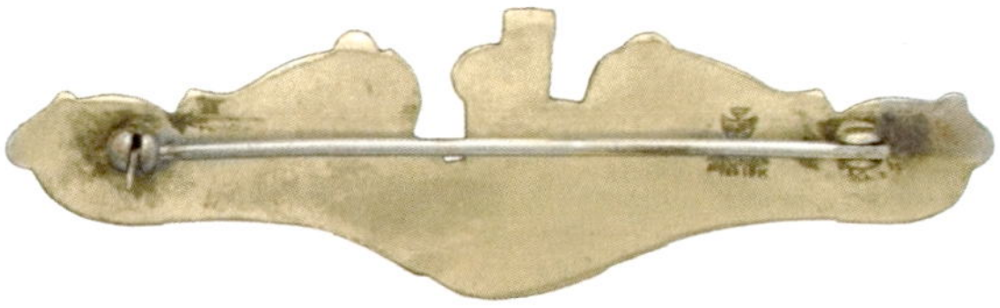
USN-1-JO.1r

USN-1-JO..2

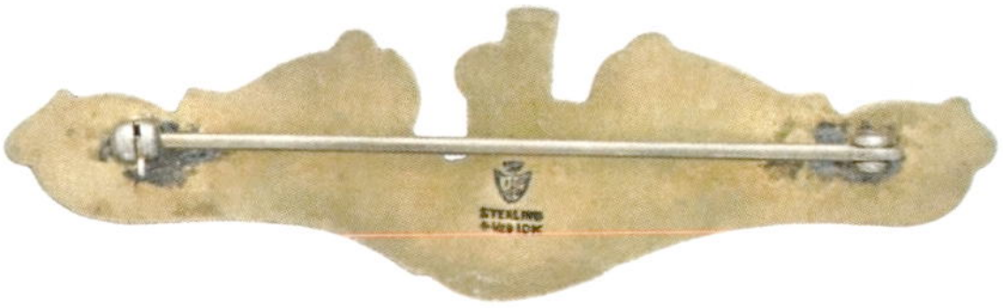
USN-1-JO.2r

Robbins Company
Attleboro, MA (R21)

USN-1-ROB.1

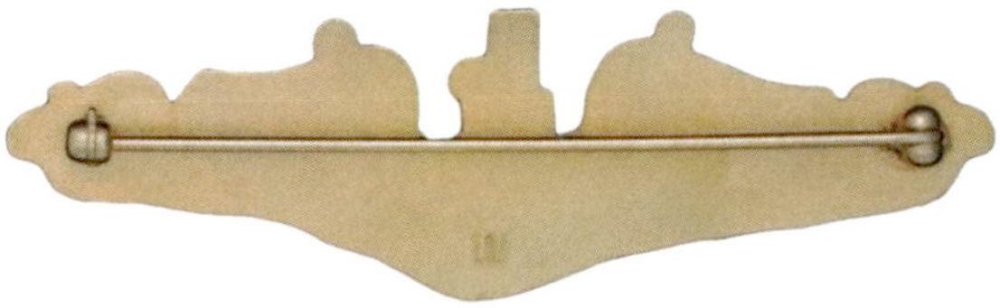
USN-1-ROB.1r

USN-1-ROB.2

USN-1-ROB.2r

USN-6-ROB.1

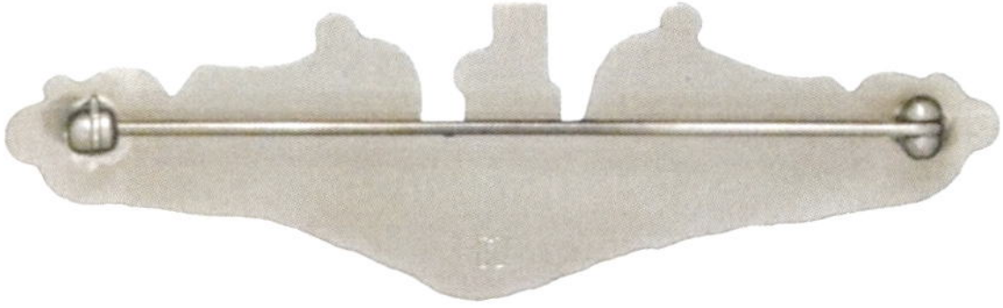

USN-6-ROB.1r

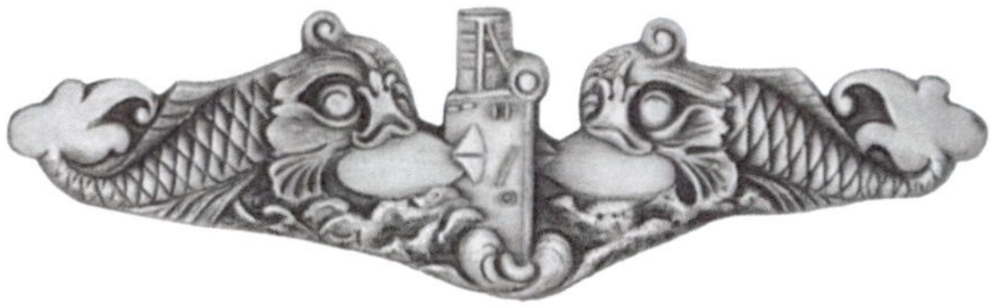

USN-6-ROB.2

USN-2-ROB.2r

Pasquale Uniform Company
San Francisco, CA.

USN-1-PASQ.1

USN-1-PASQ.1r

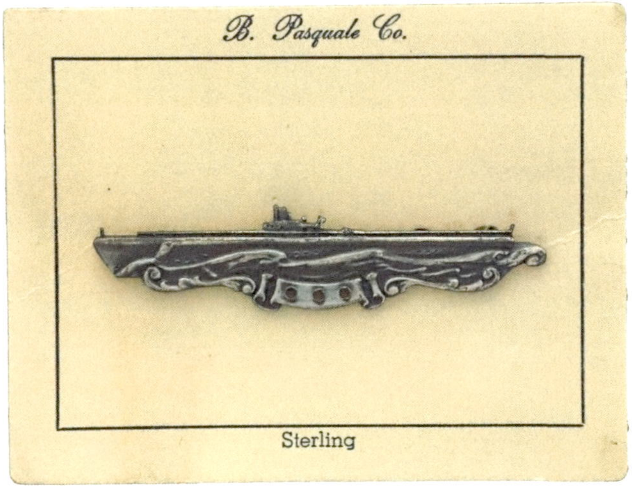

US-3-PASQ.1

USN-3-PASQ.1r

Russell Uniform Co.
New York

USN-6-RUS.1

Sheridan
Perth, Western Australia

USN-1-SHER.1

USN-1-SHER.1r

USN-3-SHER.1

USN-3-SHER.1r

STSILVER - Australia

USN-3-STSILVER.1

USN-3-STSILVER.1r

USN-3-STSILVER.2

USN-3-STSILVER.2r

Vanguard Military Equip. Co/Vanguard Insignia Co.
(1V, V21)

USN-1-VAN.1.1

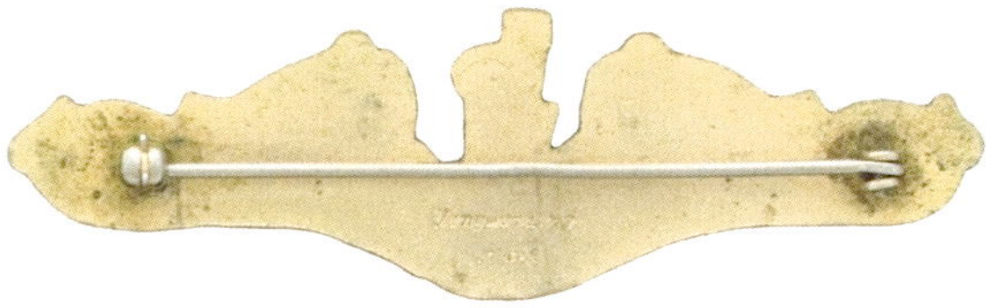

USN-1-VAN.1.1r

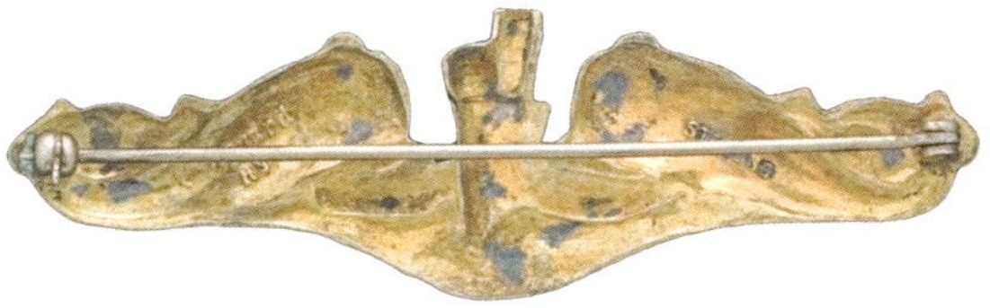

USN-1-VAN.1.2r

USN-1-VAN.1.3r

USN-1-VAN.2.1

USN-1-VAN.2.1r

USN-3-VAN-1.1

USN-3-VAN-1.1r

USN-3-VAN.2.1

USN-3-VAN.2.1r

USN-3-VAN.2.2r

USN-3-VAN.2.3r

USN-6-VAN.1.1

USN-6-VAN.1.1r

USN-6-VAN.1.2r

USN-6-VAN.1.3r

USN-6-VAN.1.4r

USN-8.1-VAN.1

USN-8.1-VAN.1r

USN-8.1-VAN.2r

Manufacturers not included with those shown are:
Bond, J.E. Caldwell, Collinson, Crest Craft, Dodge, H.L.I. Lordship, Luxenberg, Saunders, Wolf & Brown, Yardley and others

VENEZUELA (VEN)

The Venezuelan Submarine Force was created when the Navy purchased and commissioned an ex-*Balao* class submarine from the United States in 1960 with a few more submarines purchased from the U.S. until the early 1970s. Between 1976 and 1978, Venezuela purchased two German Type 209/1300 diesel electric submarines to replace their aging fleet submarines. These two submarines remain in service (at least on paper) to the current day. Venezuela also currently operates a diver transport mini submarine.

The Venezuelan submarine qualification insignia was introduced in an unofficial capacity sometime in 1961, shortly after the commissioning of their first boat. According to sources, the first badge used was very similar to the U.S. badge, however, by 1965, the Venezuelan badge design as we know it had emerged. This design was loosely based on the US badge with two inward-facing heraldic dolphins flanking a slightly port-facing submarine sail with planes and periscopes, the national emblem in the center just below the sail. This badge was issued in gilt metal for officers (VEN-1) and in silver metal for enlisted crew members (VEN-2). These badges were typically clutch back and are worn on the upper left chest of all uniforms, above the ribbon bar if present. On 27 November 1991, Ministerial Resolution No. DG 6066 was issued officially implementing these badges.

VEN-1

VEN-2

Shortly after the 1991 official authorization of the VEN-1 and VEN-2 badges, according to sources around the end of 1992 or the beginning of 1993, a new badge was approved for submarine commanding officers. Taking advantage of having a former naval officer serving as the country's Minister of Defense, officials in the submarine squadron were able to push through design approval. This badge reflected the same design as the officer's badge but has seven stars in an arc over the submarine sail (VEN-3).

VEN-3

In 2004 Ministerial Resolution No. DG 26269 introduced a new badge. Another badge with the same gold officer's badge design was fashioned, this one with a single large star surrounded by a wreath above the submarine's sail. This badge was presented to submarine squadron commanders (VEN-4). Many collectors have seen a badge that has just a single large star above the sail without the wreath. This badge is completely unauthorized and is considered a fake.

VEN-4

The national emblem on the badges between the 1960s and 2006 had Simón Bolívar's horse on the bottom galloping to the right and looking back to the left. Following the accession of the Chavez government in 2006, the national emblem was modified in a few ways, most noticeably Simón Bolívar's horse is now galloping to the left. Slightly less noticeably, a machete and bow and arrows were added to the weapons in the top right yellow field of the emblem. These changes are reflected in all submarine badges made after 2006; the enlisted badge (VEN-5), the officer's badge (VEN-6), the commanding officer's badge (VEN-7), and the squadron commander's badge (VEN-8).

VEN-5

VEN-6

VEN-7

VEN-8

The 2004 Resolution also provided for "Honorary Submariner" badges to be issued. These were either the officer or enlisted personnel badges presented to individuals who provided invaluable services to the submarine community. When presented, this badge would be worn on the right side of the uniform above the pocket to distinguish the individual from a qualified submariner.

Beginning in the mid-2000s, patches were authorized for use on submarine coveralls, with gold thread on a dark blue background (VEN-5C & VEN-8C) and black thread on a camo green background (VEN-7C) patches known.

VEN-5C

VEN-8C

VEN-7C

VIETNAM (VIE)

The Vietnam People's Navy (VPN) is one of the more recent navies to acquire submarines, but they jumped in with both feet. Purchasing six *Kilo*-class submarines from Russia in 2009, the VPN's 189th Submarine Brigade was created when the six submarines were commissioned on 29 May 2013.

Shortly after creation of the submarine service, submarine badges were introduced. The basic design for the badges is a gilt port-facing submarine profile resting on the arms of a ships anchor, a star on the shank, with a background of stylized waves in blue and gold. There are two badges, one with *Thuyền Trưởng* or "Captain" on the stock of the anchor in red letters issued to submarine Commanding Officers (VIE-1), and the other with *Hải Quân Việt Nam*, or "Vietnam Navy" issued to the officers (VIE-2). They have horizontal pin attachment and were worn on the left chest above the pocket or ribbon bar. There is no known miniature metal badge or patch. Enlisted personnel were not issued badges.

VIE-1

VIE-2

These badges were in use until 2022 when all specialty badges were eliminated and replaced with a basic service badge that was issued to all members of the VPN (VIE-3). At this time, there is no plan to reinstate specialty badges.

VIE-3

YUGOSLAVIA (YUG)

The Kingdom of Serbs, Croats, and Slovenes was formed in 1918 near the end of WWI. Colloquially known as the Kingdom of Yugoslavia, they made their first foray into submarines with the purchase and commissioning of two British *L*-class submarines on 1 April 1928. The *Hrabri* (Brave) and the *Nebojša* (Fearless) were accepted into the Yugoslav fleet with great fanfare… but without any existing submarine distinctives. A year later, around the same time that King Alexander officially changed the name of the country to the Kingdom of Yugoslavia, it was announced that two French-built submarines would be joining the fleet, and with that announcement, submarine insignias were also introduced.

Two versions of the Yugoslav submarine insignia were ordered at the same time. The first version was for the crews of the *Hrabri* and the *Nebojša.* A silver port-facing *Hrabri*-class submarine profile riding stylized waves laying across a gilt wreath of olive leaves and branches, bound at the bottom with a bow, and topped with a pierced red enameled royal crown (YUG-1). There appears to be at least two versions of this badge, one with top and bottom screw attachment points, and one with two vertical pins.

YUG-1

YUG-1r

The second version of the first insignia was issued to the crews of the new French-built submarines *Osvetnik* (Avenger) and *Smeli* (Daring). Very similar to the first badge, but with some marked differences, the silver port-facing *Osvetnik*-class submarine rides alone and not on stylized waves, but with prominent deck guns not featured on the first badge. The gilt wreath is merely wrapped at the bottom without a bow, and is slightly thinner and smaller, but is topped by the same pierced red enamel royal crown (YUG-2) and is typically found with two vertical pins for attachment.

YUG-2

YUG-2r

These badges were worn by all personnel on the left chest, and were manufactured by the Zagreb jeweler, Griesbach and Knaus, with most known examples being numbered. There are conflicting reports as to whether these badges were considered to be officially sanctioned or not, however, though no official approval for them has yet been identified, they were presented in official ceremonies. Any controversy regarding official status was removed by the next badge.

On 10 December 1937, King Peter II of Yugoslavia issued royal decree 29076 "To introduce the 'Insignia of the Submariner' as a visible recognition for the hard and dangerous service on Kr. Submarines." This insignia was to be made of 900/1000 purity silver, a laurel and oak leaf wreath with the royal crown on top and the official state seal at the bottom, a fouled naval anchor standing vertically in the middle and a port-facing submarine profile overlying the wreath and anchor horizontally in the center (YUG-3). The majority of these badges were made by the Belgrade mint, "Kovnica" and were numbered on the back of the anchor and the submarine (YUG-3-K.1 & 3-K.2). Examples exist by Griesbach and Knaus, but in fewer numbers (YUG-3-GK). Multiple modern copies of the YUG-3 have been created.

YUG-3-K.1

YUG-3-K.1r

YUG-3-K.2

YUG-3-K.2r

YUG-3-GK

These badges were only in use until April 1941 when the Kingdom of Yugoslavia was occupied and divided by the Axis powers and their allies. It was not until 1945 following the end of the war and the formation of the Socialist Federal Republic of Yugoslavia that the country operated submarines again, and even longer until new submarine badges were introduced.

The next generation of Yugoslavian submarine badges was implemented in 1961. A patinated silver metal diamond-shaped and slightly convex shield features a port-facing stylized submarine profile riding waves, with the Communist red star in enamel at the top, and the arms, shanks and rope of two naval anchors visible behind (YUG-4). This badge was issued only to officers and was worn on the left chest of all uniforms and features a single screw back attachment point. This badge was the only official submarine qualification badge worn in the Yugoslavian navy until 1992. When the recipient finished his service, it was expected that the badge be returned to the navy. According to sources, these badges were rarely returned. Gold versions of this badge have been manufactured that sources say were issued to VIPs, though many more were created for the collector's market and are not officially sanctioned badges. Multiple copies and fakes of the silver badge were also produced.

YUG-4

In 1970, two more badges were introduced, also only for officers. The design of these badges is a silver laurel leaf wreath bound at the bottom with a red enamel star at the top, a port-facing submarine horizontal across the middle above a gold rising sun and blue water with a Hall's anchor. These were announced as "Senior Submariner" badges, with five years aboard submarines earning the badge with a silver submarine (YUG-5) and ten years earning the badge with the gold submarine (YUG-6). Both badges were slightly convex with a single screw attachment point and a small stabilizing pin at the top, the badges created from 1970-1974 having the 0.900 silver jeweler's hallmark and a serial number on the back. After 1974 the material changed to tombak from silver, the stabilizing pin went away, and the numbers doubled in size. These badges were also used until 1992. The recipients of these badges were allowed to keep them. Fake versions of these badges exist, including one with a gold submarine <u>and</u> a gold wreath.

YUG -5

YUG-6

After the dissolution of the Socialist Federal Republic of Yugoslavia that occurred from 1991 to 1992, the submarines were all shifted to Montenegro under control of the new Federal Republic of Yugoslavia (FRY) government, but all previously used insignias were banned. Though the submarine force remained active, there is no record of any officially sanctioned submarine insignia being worn by Yugoslavian submariners after 1992 despite badges existing with claims to be from that time period. In January of 1998, the FRY Ministry of Defense approved a Yugoslavian Armed Forces guideline for all plaques and insignias. Article 26 of this guideline spelled out the criteria for proposed new submarine insignia for officers and non-commissioned officers.

This new design was a double-headed eagle perched on the arms of an admiralty anchor, a blue shield under the stock with a starboard-facing gold submarine in the center, a Hall's anchor in the middle of the submarine and heraldic dolphins astride the submarine conning tower. According to the guidelines, this badge was to be made of "900/000" silver and was to include a registration number and maker's mark on the back. Gilt colored for officers (YUG-7-UA) and silver colored for non-commissioned officers (YUG-8-UA), unfortunately, this badge design was never produced in an authorized capacity, and only officially existed on the pages of the guideline, and on qualification certificates awarded to FRY submariners after 1998. While physical examples exist, these are fantasy pieces produced by the Serbian Submarine Association and they are neither numbered nor hallmarked, are cast, and are not the high quality that would be expected of 900/1000 silver.

YUG-7-UA

YUG-8-UA

There were plaques using the 1998 design that were officially issued to the command staffs of the two existing submarines (YUG-PLQ-1 & YUG-PLQ-2). An important aspect of these plaques is, as issued, the eagle portion of each plaque was very fragile and often broke. This has resulted in the center metal anchor and submarine section being sold as badges – which they are not. They are simply the less fragile center of the plaque.

YUG-PLQ-1

YUG-PLQ-2

The FRY dissolved in 2003 becoming the State Union of Serbia and Montenegro. When those two countries declared independence in 2006 the last Yugoslavian submarine was decommissioned, so no new badges are available, and no known submarines or badges are forthcoming at the time of publishing from either Serbia or Montenegro.

REFERENCES

Albania

William Crosby

Commander Hans Chr. Lauritzen, RNoN, Ret.

Algeria

Commander Hans Chr. Lauritzen, RNoN, Ret.

Argentina

Captain Gustavo Conde, Ret.

Capitan de Navio VGM Gustavo Enrique Piccardo

Justo Manuel A. Rodriguez

Australia

T.J. Carter

Andrew Chaney

David John Payne

Jeff Williams

Austria-Hungary

William Crosby

Bangladesh

Rear Admiral (Upper Half) Masud Iqbal

Bolivia

William Crosby

Brazil

Raimundo Rodrigues

Bulgaria

Leonid Kanterman

Naval Museum Varna

Canada

George Quigley

Rick Kennedy

Chile

Piero Fagandini González
Rear Admiral, Submariner, Staff Officer, Academy Professor, Full Member of the Academy of Naval and Maritime History of Chile

Colombia

Captain Antonio Javier Espitia Porras
Naval Attaché Embassy of Colombia in the United States of America

Capitán de Corbeta (Lieutenant Commander) Mario A Rivera Vargas

Juan Carlos Gutiérrez Pérez

Cuba

Chief Engineer Guillermo Valdés Correoso, Ret.

Denmark

Torkil Nissen

Ecuador

Manuel Enrique Muñoz Macias

Captain Gerardo Vite

Estonia

David John Payne

Mike Brookins

Finland

Riitta Blomgren
Curator, Military Museum Vantaa

France

Eric Gragnic

Germany

Walter Kryskiw

Michael Setzer

Greece

Commander Athanasios Katsikis, Hellenic Navy, Ret.

Christy Emilio Ioannidou Author – Naval History Researcher Hellenic Institute for Maritime History

Commander George Apidianakis, GRC
Hellenic Navy General Staff
Naval History Department

Lieutenant Commander Thanos Papadimitropoulos Hellenic Navy – Naval History Department

Theofanis Vlachos
History Archive Curator
Athens War Museum

Zino Stef

Indonesia

Ivan Abraham

Iran

Commander Hans Chr. Lauritzen, RNoN, Ret.

Israel

Captain Eyal Ben Zion, Ret.

Leonid Kanterman

Italy

Massimo Agatti

Edoardo Maria Dallavalle
Curator Italian Online Naval Museum

Alberto Menichetti

Japan

Nick Komiya

Akira Komiya

Malaysia

Syed Muhammad Shafiq

Netherlands

Peter Van de Put

Nigeria

William Crosby

North Korea

Steve Jones

Norway

Commander Hans Chr. Lauritzen, RNoN, Ret.

Pakistan

Rear Admiral Syed Zain Zulfiquar, Ret.

Peru

Ts2 Ars. (SS) José Collazos

Poland

Adam Tomasz Krzykowski

Portugal

João DaCruz Gonçalves

Romania

Mike Brookins

Saudi Arabia

William Crosby

Singapore

Junhao Gregory Wong

South Africa

Eric Fraser

John Lamont

South Korea

Captain Il Choi, Submarine Analyst, Ret.

Soviet Union/Russia

John Dyuk

Leonid Kanterman

Sergei Kazeka

Spain

Naval Museum of Cartegena, Spain

Jose (Pepe) Martinez

Sweden

Swedish National Maritime Museum, specifically,
Margrit Wettstein, PhD Curator, Collections Unit, Naval Museum
& Fredrik Kämpe
Swedish Naval Museum and PhD student at Stockholm University

Lars Nordenberg
Chairman of the associations:
Submarine organization Nordkaparen Gothenburg
Fishing Museum at Hönö.

Pär Flyghed

Peter Olausson

Taiwan (Republic of China)

Commander William Lu, ROCN

Master Chief Petty Officer Tso Te Wen, ROCN

Thailand

Commander Pakphum Phairotchananan

Turkey

İnanç Murat Gamsiz

Ozkan Kaya

Emre Cabbaroglu

Ukraine

John Dyuk

United Kingdom

Owen Pate

United States

Joe Dietrich

Capt. David O. Hill

David Jones

Commander Steve Jones, USN, Ret.

Mike (Lee) Lacey

Bill Miller

Captain Don Walsh, USN, PhD, USN Deep Submersible
Pilot #1, Ret.

Russel Wilson

Venezuela

Capitán de Corbeta (Lieutenant Commander) Guido Villamil, Ret.

Ramon Rivero

Viet Nam

Văn Thông Trần

Senior Lieutenant Minh Chiến Trần

Yugoslavia

Dusan Babac

Jovan Mara

BIBLIOGRAPHY

Babac, Dusan M. *Military Insignia in Serbia, 1845-1945*. Official Gazette Publishing. 2014.

Dimitriev, Arcady Anatolevichem. *Badges and Medals of Russian Submarines, History of Creation, Traditions and Development of Symbols, 1909-2002.* Central Design Bureau "RUBIN" Printing Office. 2002.

Fagandini González, Piero, (Rear Admiral). *Brief Review on the Qualification Insignia of the Submarine Specialty of the Chilean Navy*. Uncopyrighted monograph, Santiago, Chile. 2022.

Gragnic, Eric. *Les Certificats De Sous-Marinier Francais, de l'origine a nos jours.* Self-published, France. 2017.

Gragnic, Eric, and Jean-René Buchbach. *Atlas des Écussons Sous-Marins Nucléaires D'Attaque Français.* Uncopyrighted monograph, France. 2016.

"History of the Spanish Submarine Weapon, Lecture Series." *Cuaderno Monográfico,* Num. 71, Madrid, 2015.

Jimenez Rattia, Eladio J.G. *The Steel Fish... More Than Forty Years of History.* Venezuelan Naval Printing Office. 2000.

Jones, David A. *US Silent Service Dolphins & Combat Insignia 1924-1945,* R. James Bender Publishing. 2001.

Lamont, John, and Eric Fraser. *South African Submarine Badges.* Uncopyrighted monograph, South Africa. 2020.

Lauritzen, Hans Christoffer. *Norwegian Dolphins: 1958-2022*. Uncopyrighted monograph, Stavanger, Norway. 2022.

Meagher, P. (n.d.) "The DBF Pin." *SubmarineSailor.com,* http://www.submarinesailor.com/history/dbfpin/dbfpin.asp#:~:text=The%20DBF%20pin%2C%20originally%20designed,qualified%20and%20served%20on%20smokeboats.

Menichetti, Alberto. "Italian Submariners' Badges, 1918-1945." *UNIFORME E ARMI*, N.4, August 1989. Translation by Massimo Agati. 2004.

Menichetti, Alberto. "The Italian Navy Long War Navigation Badges." *UNIFORME E ARMI*, N.41, July 1994. Translation by Massimo Agati. 2004.

Menichetti, Alberto. "The Honour Insignia for the Navy Assault Detachment Personnel." *UNIFORME E ARMI*, N.52, June 1995. Translation by Massimo Agati. 2004.

Petrov, Panayotov, Zhechev. *Breast Decorations and Badges Awarded to Bulgarian Submarine Men.* BM Trade. 2009.

Phillips, Allen. *A Guide to Submarine Insignia.* (Original Title: *Iron Men, Steel Boats, Metal Badges*.) Dolphin Glass Press. 1994.

Prichard, Pete. *Submarine Badges and Insignia of the World – An Illustrated Guide for Collectors*. Schiffer Military/Aviation History. 1997.

Setzer, Michael. "Submarines and Submariners: From the Cigar Box to the Jewelry Display Cabinet." *Auftauchen!* August-October 2022, pp. 8-10.

"Submarine Dolphins." *Navalunderseamuseum.org,* 19 January 2017, https://navalunderseamuseum.org/submarine-dolphins/?fbclid=IwAR12H6sY7eW5Bv8gCL7qJFXdDqnathvlCeuxU5Fv7hpolYQrs50P0WuShGA

Thornton, W.M. *Submarine Insignia & Submarine Services of the World.* Pen & Sword Books Ltd. 1997.

www.ingramcontent.com/pod-product-compliance
Ingram Content Group UK Ltd.
Pitfield, Milton Keynes, MK11 3LW, UK
UKRC031012141026
15625UKWH00082B/659

9798218866952

9 798218 866952